Industrial Brushless Servomotors

Industrial Brushless Servomotors

Peter Moreton

Newnes

OXFORD AUCKLAND BOSTON JOHANNESBURG MELBOURNE NEW DELHI

Newnes
An imprint of Butterworth-Heinemann
Linacre House, Jordan Hill, Oxford OX2 8DP
225 Wildwood Avenue, Woburn, MA 01801-2041
A division of Reed Educational and Professional Publishing Ltd

& A Member of the Reed Elsevier plc group

First published 2000

British Library Cataloguing in Publication Data
A catalogue record for this book is available from the British Library.

ISBN 0 7506 3931 8

Library of Congress Cataloguing in Publication Data
A catalogue record for this book is available from the Library of Congress.

Transferred to digital printing 2005

Typeset by Jayvee, Trivandrum, India
Printed and bound by Antony Rowe Ltd, Eastbourne

Contents

Preface

The industrial brushless servomotor has developed through a remarkable combination of mechanical, electrical, power electronic and microelectronic technologies, and both the operation and application of the motor rely on many interdependent factors. I have tried to cover the fundamentals of the subject in a logical manner, taking a step-by-step approach, describing first the construction of the brushless machine itself and how it works, second, how the current is supplied, third, how the motor behaves when it is loaded and finally how it is rated and selected for a particular duty. The book covers the important motor and load characteristics which affect the design of the control system, but does not include a detailed treatment of control techniques which are well described elsewhere.

The first chapter is devoted to a brief review of the brushed, permanent magnet motor. This allows the early introduction to the book of some basic groundwork using what is perhaps a more familiar machine, and also allows a clearer comparison to be made with the brushless type later on. Throughout I have been aware of the needs of engineers and students with no previous knowledge of how brushed or brushless motors work, and so both forms are explained from first principles. Theoretical analysis is developed in relation to practical examples, and rules of thumb are suggested wherever possible. Any equations for motor rating and selection are simple enough for numerical results to be found using a calculator or spreadsheet. My hope is that this publication will be of some help to those who are already using brushless motors in servomechanisms, as well as to those who are studying the electrical and mechanical properties which are involved.

The practical nature of this book has been made possible by the generous supply of technical advice from the members of staff of SEM Ltd. I wish to acknowledge a debt of gratitude to Paul Newall for his constant support and for the many hours of his time taken up by our discussions, and also to Van Hamlin and Omar Benzaid for their readily given advice and practical help. I am also indebted to several members of staff of the University of Bristol, and wish to acknowledge here the help given by two in particular. Duncan Grant suggested the basic idea for the book and followed through with advice and encouragement from start to finish. I am also extremely grateful to have had the very willing help, particularly with the systematic solution of quartic equations, of Gordon Reece of the Department of Engineering Mathematics. Finally, I would like to give a special thanks to Paul Prater of Lewis Berl Automation.

Acknowledgement

The various photographs were kindly supplied by the following companies:

SEM Ltd, Kangley Bridge Road, London SE 26 5AS, UK

Parker Hannifin GmbH, Hauser Division, Robert-Bosch-Str. 22, 77656 Offenburg, Germany.

List of units

Unit symbol	Name
A	ampere
°C	degree centigrade
H	henry
mH	millihenry
Hz	hertz
J	joule
kg	kilogram
m	metre
mm	millimetre
A-turn	ampere-turn
H/m	henry per metre
kg m^2	kilogram-square metre
N	newton
Nm	newton metre
rad	radian
μrad/Nm	microradian per newton metre
s	second
ms	millisecond
m/s	metre per second
Nm/rad	newton metre per radian
Nm/rad s^{-1}	newton metre per radian per second
T	tesla
V	volt
V/rad s^{-1}	volt per radian per second
W	watt
°C/W	degree centigrade per watt
Wb	weber
Ω	ohm

List of symbols

Symbol	Definition	Units
AC	alternating current	
B	magnetic flux density	T
C	compliance	μrad/Nm
c_p	profile constant	
d	screw pitch	m
D	damping constant	Nm/rad s^{-1}
DC	direct current	
e	base of the natural logarithm	
E	electromotive force (emf)	V
F	force	N
G	gear ratio	
H	magnetic field intensity	A/m
i	instantaneous current	A
I	current	A
I_{rms}	root-mean-square current	A
I_S	continuous stall current	A
j	imaginary operator $\sqrt{-1}$	
J	moment of inertia	kg m^2
J_m	motor moment of inertia	kg m^2
J_L	load moment of inertia	kg m^2
J_r	ratio of load to motor moments of inertia	
K_E	voltage constant	V/rad s^{-1}
K_T	torque constant	Nm/rad s^{-1}
L	inductance	H
L_{LL}	brushless motor inductance, line to line	
l	length	m
m	mass	kg
N	number of turns	
N_s	number of turns on a sinusoidal winding	

P	power	W
P_{sp}	speed-sensitive loss	W
R	resistance	Ω
R_{th}	thermal resistance	°C/W
$R_{\text{th}}\tau_{\text{m}}$	motor rating coefficient	°C ms/W
R_{LL}	brushless motor resistance, line to line	Ω
p	profile distribution factor	
r	radius	m
SI	international system of units	
s	Laplace operator $s = \sigma + j\omega$	s^{-1}
T	torque	Nm
T_{L}	load torque	Nm
T_{S}	continuous stall torque	Nm
T_{soac}	continuous rated torque	Nm
T_{rms}	required torque	Nm
t	time	s
t_{p}	duty operation time	s
t'	duty cycle period	s
V	circuit input voltage	V
v	velocity	m/s
x	linear displacement	m
ϵ	energy	J
Φ	magnetic flux	Wb
ϕ	stator angle of sinewave motor conductors	rad
μ	permeability	H/m
Θ	temperature	°C
Θ_0	ambient temperature	°C
Θ_{SS}	steady-state winding temperature	°C
Θ_{pk}	peak, winding ripple temperature above Θ_0	°C
Θ_{av}	average, winding temperature above Θ_0	°C
Θ_{min}	minimum, winding ripple temperature above Θ_0	°C
θ	angular displacement	rad or°
θ_{p}	angle of load rotation	rad
σ	real part of Laplace operator s	
τ_{e}	electrical time constant of motor	s
τ_{m}	mechanical time constant of motor	s

τ_M	mechanical time constant of motor and load	s
τ_{th}	thermal time constant of motor	s
τ_w	thermal time constant of motor winding	s
ω	angular velocity	rad/s
ω_m	motor velocity	rad/s
ω_L	load velocity	rad/s
ω'	constant velocity of motor	rad/s
ω_c	constant velocity of load	rad/s

CHAPTER 1

BRUSHED DC MOTORS

1.1 Introduction

Industrial brushless servomotors can be divided into two main types. One operates in a similar way to the three-phase synchronous motor and the other is a relatively simple development of the brushed DC motor. Both types of brushless motor have the same sort of construction and have an identical physical appearance. Both have many characteristics similar to those of a permanent magnet brushed DC motor, and both are operated from a source of direct current. A review of the features of the permanent magnet brushed motor is therefore a convenient first step in the approach to the brushless type. In this first chapter, the relationships between the supply voltage, current, speed and torque of the brushed motor are developed from fundamental electromagnetic principles. Attention is also given to the factors controlling the steady-state speed of the unloaded motor.

The later part of the chapter is devoted to the question of DC motor rating. Only the basic ideas are covered at this stage, in preparation for the more detailed treatment in Chapter 5. The power losses which lead to motor temperature rise are identified, and the main factors affecting the final steady-state temperature are explained for both continuous and intermittent operations of the motor. The scope of this chapter is confined to cases where the losses during periods of speed change are insignificant in comparison to those generated during the periods of constant motor speed.

1.2   Operational principles

Motor construction

Figure 1.1 shows the essential parts of a rudimentary permanent magnet DC motor. Two conductors are connected in series to form a winding with one turn. The winding has a depth l and width $2r$ metres and is mounted between the poles of a permanent magnet. The winding is free to rotate about the dotted axis and its ends are connected to a DC source through sliding contacts to form a circuit carrying current I A. The main diagram is drawn for the moment when the conductors are passing the centre of the poles.

The contacts allow the direction of current in the winding to reverse as it moves through the vertical position, ensuring that the direction of flow through the conductors is always the same relative to the direction of the magnetic field. In other words, it does not matter in the diagram which side of the winding is to the left or right when we look at how torque is produced.

Torque production

The torque produced by the motor in Figure 1.1 is the result of the interaction between the magnetic field and the current-carrying conductors. The force acting on each conductor is shown as F. Some simple magnetic principles are involved in the evaluation of the torque.

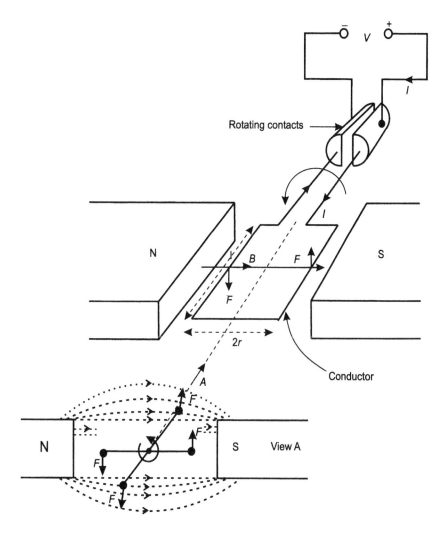

Figure 1.1
Principle of the permanent magnet brushed DC motor

Magnetic flux Φ

The amount of *magnetic flux* in a magnetic field tells us how much magnetism is present. By itself, it does not give the strength of the field. The flux may be represented by lines drawn between the poles of the magnet and in the old British system the unit of flux was, in fact, the line. In the SI system the unit is the *weber*, denoted by Wb, where one weber is equivalent to 10 lines in the old system.

Magnetic flux density B

As its name suggests, the term *magnetic flux density* describes the concentration of the magnetic field. The SI unit of magnetic flux density is the *tesla*, denoted by T, where a tesla is equal to one weber per square metre.

The force on a conductor

When a conductor of length l, carrying a current I, is placed in a magnetic field of uniform flux density B, it is found that the conductor is acted on by a force which is at right angles to both the field and the conductor. The force is greatest when the conductor and field are also at right angles, as in Figure 1.1. In this case, the force is given by

$$F = BlI \text{ (N)}$$

The unit of force is the *newton*, denoted as N. The direction of F can be found by the 'left-hand motor rule'. This states that the thumb of the left hand points in the direction of the force, if the first finger of the hand is pointed in the direction of the field and the second finger in the direction of the current.

Torque

Force F acts on each conductor of the winding shown in Figure 1.1. The torque produced at each conductor is

$$T = Fr \text{ (N m)}$$

The unit of torque is the *newton metre*, denoted as Nm. The radius of action of F around the axis falls as the winding moves away from the horizontal position, reducing the torque. In the figure, the winding lies in a plane between the centres of the flat poles of the magnet, where B is greatest. With such a pole shape the flux will be less dense at other winding positions, reducing the torque still further.

Figure 1.2 shows three practical DC motors with the circular type of pole faces shown in Figure 1.3. These give a substantially radial and uniform pattern to the flux so that B and T remain constant in the ideal case. The winding has a number of turns, with the conductors distributed in slots (not shown in cross-section) around a cylindrical iron carrier, or rotor. For simplicity, the cross-section shows only seven turns, each with two conductors arranged diametrically. The current directions are shown by the use of a cross and a dot for current flowing into and out of the paper respectively. The turns of the rotor winding are connected to the segments of a commutator which rotates between spring-loaded brushes. The current in each turn of the winding reverses each time the turn passes the brush axis, and the pattern of crosses and dots in Figure 1.3 will be the same for any rotor position. The reversals give a rectangular AC waveform to the current in the individual turns of the motor winding. Only the brushes carry a unidirectional current.

Figure 1.2
Permanent magnet DC motors

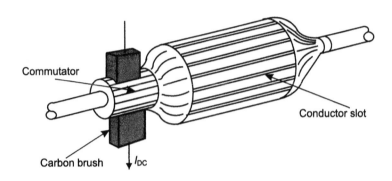

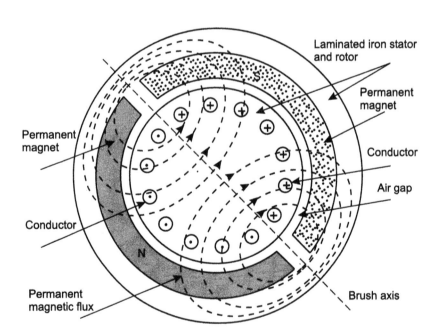

Figure 1.3
Cross-section and rotor of a two-pole, permanent magnet DC motor

For a winding with N turns, there are $2N$ conductors. The finish of each turn is joined to the start of its neighbour at a segment of the commutator. Two circuits of $N/2$ turns appear in parallel between a pair of brushes which touch segments at opposite sides of the commutator, and so each of the $2N$ conductors carries a current of $I/2$. The combined torque is

$$T = NBllr$$

Assuming that the poles of the motor in Figure 1.3 are the same length l (into the paper) as the conductors, we can write the flux density around the face of each pole in terms of webers per square metre as $\Phi/\pi rl$. The torque expression for the two-pole motor with one winding of N turns becomes

$$T = \frac{N\Phi I}{\pi}$$

The torque constant

For any given motor, the only variable in the last expression is the current I. The torque can be expressed as

$$T = K_T I$$

K_T is the *torque constant*, expressed in Nm/A. It is one of the most important constants in the motor specification.

Motor speed

When the voltage is switched on to an unloaded DC motor, the rotor speed rises from zero and quickly reaches a 'no-load' terminal value. The normal losses associated with the DC motor itself would not be enough to prevent the speed from rising to a point very much higher than the no-load value, and the question arises of how the limit in speed occurs. To answer, we must look at a second aspect of the behaviour of a moving conductor in a magnetic field.

Voltage generation

Figure 1.4 shows a conductor of length *l* which is being moved with velocity *v* metres per second (m/s) across and at right angles to a uniform magnetic field of density *B*. As the conductor moves across the field, a voltage known as the electromotive force or *emf* will be generated along its length equal to

$$E = Blv \ (\text{V})$$

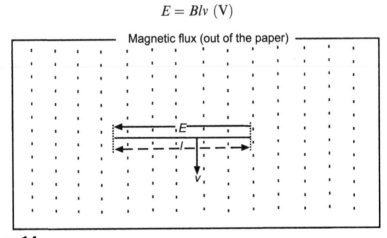

Figure 1.4
Conductor moving across a magnetic field

In the 'right-hand generator rule', the second finger points in the direction of *E* if the forefinger is pointed in the direction of the field and the thumb in the direction of movement. The rotor of a two-pole motor with a winding of *N* turns has 2*N* conductors, and there are always two parallel paths of *N* conductors connected in series between the brushes. The conductors travel at a speed of *ωr*, where *ω* is the *angular velocity* expressed in radians per second or rad/s. The total voltage induced between the start and finish of the winding is therefore

$$E = NBlv$$

Substituting for *B* as before gives

$$E = \frac{N\Phi\omega}{\pi}$$

The voltage constant

In the last equation above, all quantities except ω are constant for any given motor and so the induced voltage is

$$E = K_E\omega$$

where K_E is the *voltage constant* expressed in volts/radian per second or V/rad s^{-1}.

K_T and K_E

Comparing the expressions above for T and E shows that for a two-pole motor with a single winding,

$$K_T = K_E = \frac{N\Phi}{\pi}$$

The equality is maintained when the number of pole pairs and a number of parallel windings are taken into account. Note that the constants have the same numerical value in SI units, but not in other systems of units.

Back emf and the terminal speed of the unloaded motor

Figure 1.5 shows a motor connected to a voltage source V_{DC}. E is generated in the direction which opposes the cause of its generation, namely the movement of the rotor. Accordingly, E acts against the applied voltage V_{DC} and is normally referred to as the *back emf*. Note that AC emfs are generated across the individual turns of the rotor winding. The emfs are commutated in the same way as the AC currents in the turns, so that the total back emf E appears as a direct voltage at the motor brushes.

If the mechanical losses due to friction and windage are ignored, steady-state conditions would be reached at a speed sufficient to make the induced voltage $K_E\omega$ equal the supply voltage V_{DC}, that is when the motor speed $\omega = V_{DC}/K_E$. In practice, the terminal speed and the induced voltage will be slightly lower to allow a small current to flow to supply the losses.

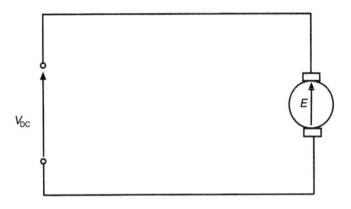

Figure 1.5
Unloaded motor at steady-state

1.3 The loaded motor at steady state

The power required to supply a torque of T Nm at a speed of ω rad/s is

$$P = T\omega \ (\text{W})$$

The unit of power is the *watt*, denoted by W. Figure 1.6 shows a DC motor connected to a load. Current flows to the motor following the application of the constant voltage V_{DC}, and the motor accelerates to a constant speed. The final steady current and speed occur when the motor output torque equals the opposing torque at the load, at which point the power output from the motor is equal to the power supplied to the load.

Steady-state characteristics

In Figure 1.7 the motor is represented by the resistance R of the rotor winding conductors, and the back emf E. The supply voltage is

$$V = RI + E$$

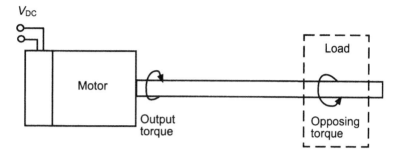

Figure 1.6
The loaded motor

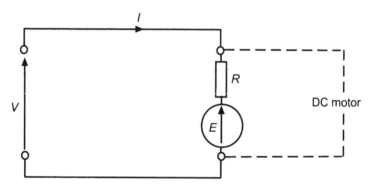

Figure 1.7
Simple equivalent circuit of a DC motor

At steady state,

$$V = RI + K_E\omega$$

The motor torque at steady state is

$$T = K_T I$$

Combining the last two equations above gives

$$\omega = \frac{V}{K_E} - \frac{TR}{K_T K_E}$$

from which we see that the speed of the permanent magnet brushed motor varies linearly with torque. The speed-torque characteristic shown in Figure 1.8 is plotted by drawing a straight line between two reference points. At the first point, when T is zero, the no-load speed is given by

$$\omega_{NL} = \frac{V}{K_E}$$

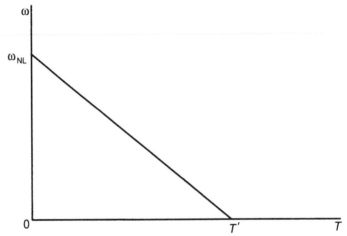

Figure 1.8
DC motor speed–torque diagram

The second reference is taken by imagining the load to increase to the point where the motor is forced to stall, making ω zero. T would then be at a theoretical maximum of

$$T' = \frac{VK_T}{R}$$

The equation for the speed–torque curve can now be written as

$$\omega = \omega_{NL} - R_{RC}T$$

where R_{RC} is the *speed regulation constant* of the motor, equal to the slope $R/K_T K_E$ of the speed–torque characteristic in Figure

1.8. The current carried by the rotor conductors rises with the motor torque. The last expression above does not take account of motor losses due to, for example, brush contact and rotor bearing friction, which in practice would cause a reduction in ω_{NL}. Figure 1.8 has been drawn for a fixed value of supply voltage. For any particular motor, a family of linear speed–torque characteristics can be drawn for a range of operating voltages. The smallest of the motors shown in Figure 1.2 is a two-pole, 24 V motor with the following constants:

$$K_T = 0.07 \text{ Nm/A}$$
$$K_E = 0.07 \text{ V/rad}$$
$$R = 0.70 \; \Omega$$

Using these constants, the no-load speed and the torque developed at the point of stall can be found at several supply voltages up to 24 V. Figure 1.9 shows the resulting characteristics. These must be applied with caution, as damage to the motor may result from the flow of high current at the low speed, high torque end.

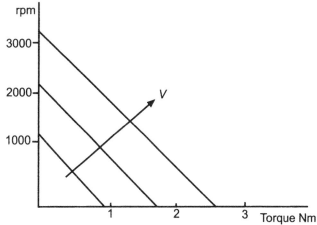

Figure 1.9
Speed versus torque at various supply voltages

Small permanent magnet DC motors have a wide range of applications such as door operators, tape drives, floor scrubbers, conveyors, as well as in small battery powered vehicles. As an example, we will take the case of an automatic sliding door which is to be driven by the small 24 V motor described above. Figure 1.10 shows the profile of door velocity. The door opens rapidly, and then crawls in readiness for its stop at the fully open position. The same action occurs in the reverse direction. For safety reasons the door closes relatively slowly, and finally crawls to its closed position. The most obvious feature of the diagram is that the motor is not required to work continuously at a constant speed, which raises the question of its rating. We should now look at DC motor ratings in general, before returning to the example.

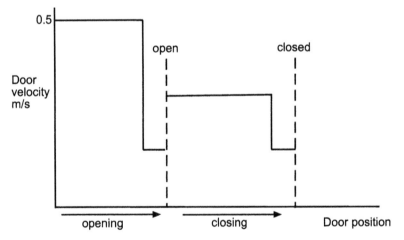

Figure 1.10
Velocity profile for an automatic sliding door

1.4   Motor rating

This section deals with ratings for continuous or intermittent motor operation, the work being broadly relevant to both the brushed and brushless motor. The intermittent operations are limited to duty cycles in which the electrical energy supplied to the motor during acceleration and deceleration may be ignored in comparison to the amount supplied over the complete cycle. Chapter 5 covers the rating of the brushless motor in more detail, and includes cases where the duty cycle demands a relatively high input of energy during periods of speed change.

Power losses

Figure 1.11 shows how the electrical power input is distributed as the DC motor performs its normal task of converting electrical energy into mechanical energy. The output power is lower than the input power by the amount of the losses, which appear mainly in the form of heat within the motor.

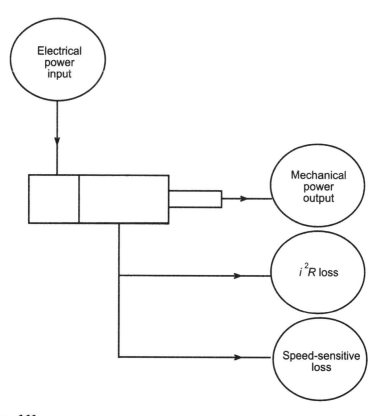

Figure 1.11
Power distribution in the DC motor

The i^2R winding loss

The flow of current I through the rotor winding resistance R results in a power loss of I^2R. Note the dependence of this loss on the motor torque $K_T I$.

Friction, windage and iron losses

As well as friction and windage, there are other effects of the physical rotation of the rotor. For example, as the rotor position changes with respect to the permanent magnetic field, flux reversals take place inside the iron core which encourage the flow of eddy currents. The consequent losses and rotor heating increase with rotor speed.

Torque loss and power loss at constant speed

The iron, friction and windage losses result in a reduction in the available output torque. The loss at constant speed is

$$T_{\mathrm{loss}} = T_{\mathrm{f}} + D\omega$$

where T_{f} is the torque due to constant friction forces, such as those produced at the rotor bearings, and D is a constant of proportionality for speed-dependent torque losses due to viscous effects such as iron losses. The constant D is known as the *damping constant* expressed as Nm/rad s^{-1}. The product of the torque loss and the motor speed is known as the *speed-sensitive loss*. Adding the i^2R loss gives the total power loss in SI units as

$$P_{\mathrm{loss}} = \omega(T_{\mathrm{f}} + D\omega) + I^2R$$

P_{loss} is the difference between the electrical power at the motor input and the mechanical power at the output shaft. Over a period of time, more energy is supplied to the motor than reaches the load. Most of the difference results in motor heating and a rise in temperature, which continues until as much heat is passed from the motor to the surrounding air as is produced internally. As there is always a designed maximum limit to the motor temperature, limits must also be set on the performance demands which lead to temperature rise. The last equation above shows that the power loss depends on motor speed and the square of the current. The current is directly related to the motor torque and we can conclude that motor speed and the square of the torque are the factors which control the temperature rise.

Continuous operation

The limits of continuous speed and torque which give rise to the maximum permissible temperature at any part of the motor are determined experimentally and plotted as a boundary on a speed–torque plane. The region to the left of the boundary is the Safe Operating Area for Continuous operation, the boundary being known as the Soac curve. Figure 1.12 shows two areas of safety, one with and one without forced air cooling. The curve takes account of the i^2R and speed-sensitive loss at all speeds and can always be used down to the stall point, unlike the basic speed–torque characteristics of Figure 1.9.

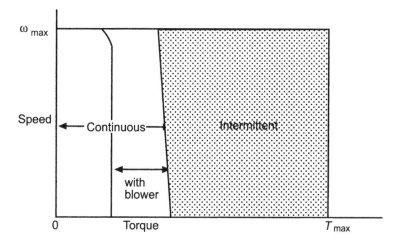

Figure 1.12
Safe operating areas on the speed–torque plane

Intermittent operation

While the area to the right of the Soac boundary may not be used for continuous running, the higher torques may still be intermittently available if the overall losses do not raise the temperature of any part of the motor above the safe limit, normally 150°C. For the brushed motor, the speed-sensitive loss is usually low in comparison to the i^2R loss. The motor losses and heating therefore depend largely on the square of the current, or effectively on the square of the motor torque. It is clearly wrong therefore to base the rating for intermittent operation on the average torque requirement. The rating on the right-hand side of the Soac boundary should be based on the root-mean-square (rms) value of the torque supplied over a complete duty cycle. Note that this applies automatically on the left-hand side, where rms and continuous torques have the same values.

At this point we may return for a moment to the example of the automatic door with the velocity profile shown in Figure 1.10. Maximum demand on the motor occurs when the door is required to open and close continuously, with the fully open periods at a minimum. The ideal motor current waveform is shown in Figure 1.13. If the current is supplied from an electronic drive, a ripple may be present on the waveform. As the same method applies for any waveform, assume for simplicity that the motor current follows the pattern shown in the figure.

The average current over the 16 second period of the cycle is

$$I_{av} = \frac{I_M(1.0 \times 2 + 0.5 \times 3 + 0.7 \times 3 + 0.5 \times 3)}{16} = 0.44I_M$$

The rms current over the same period is

$$I_{rms} = \sqrt{\frac{I^2_M(1.0^2 \times 2 + 0.5^2 \times 3 + 0.7^2 \times 3 + 0.5^2 \times 3)}{16}} = 0.56I_M$$

It is now clear that extra i^2R losses will be produced as a result of the intermittent nature of the load. The motor must be able to accept an rms current which is greater than the duty cycle average by the factor 0.56/0.44, or 1.3.

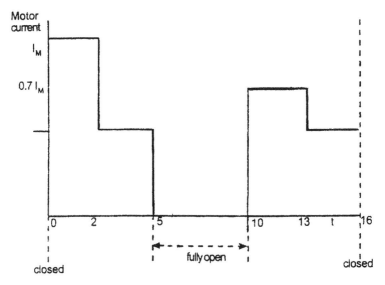

Figure 1.13
Maximum demand on door operator

The form factor

Although the above example is for one particular current waveform, the same arguments for motor rating would apply for any other waveforms. As much as possible, ratings should take into account the waveform shape defined by the term

$$\text{form factor} = \frac{I_{\text{rms}}}{I_{\text{av}}}$$

In practice, the form factor depends on a number of variables and is not always a simple, constant value.

Motor temperature

When the motor runs continuously at a fixed speed, its temperature gradually rises towards a steady-state value.

When the operation is intermittent, a ripple occurs in the plot of temperature against time. The evaluation of the temperatures relies on the use of two important motor constants.

Thermal resistance and thermal time constant

Figure 1.14 shows a rise in motor temperature for continuous operation at a constant load, from the ambient value of Θ_0 to the final steady-state value Θ_{ss}. The final temperature rise in degrees centigrade (above ambient) is

$$(\Theta_{ss} - \Theta_0) = R_{th} P_{loss} \ (^{\circ}\text{C})$$

where P_{loss} is the constant power loss at temperature Θ_{ss} and R_{th} is the *thermal resistance* in $^{\circ}\text{C}/\text{W}$. R_{th} is usually quoted as the value of thermal resistance from the hottest part, normally the rotor winding, to the air surrounding the motor case. In Figure 1.14, Θ_{ss} is therefore the final temperature of the winding.

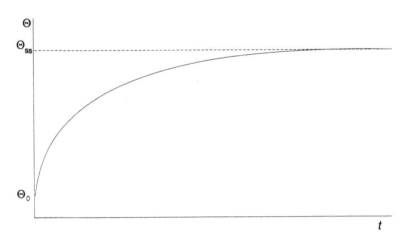

Figure 1.14
Motor temperature rise at a constant load

If the curve is assumed to rise exponentially towards Θ_{ss}, the temperature at time t is

$$\Theta = \Theta_0 + (\Theta_{ss} - \Theta_0(1 - e^{-t/\tau_{th}}))$$

where τ_{th} is the *thermal time constant* of the motor, normally given in minutes on the motor specification. The magnitude of the time constant is a measure of how slowly the temperature rises to the steady-state value. The value of τ_{th} is normally quoted for the main mass, which for brushed motors is taken to be the rotor as a whole. Note particularly that the temperature curve has the overall rate of rise of the rotor temperature, but terminates at the final value of the winding temperature.

Winding temperature ripple

When the motor runs on a duty cycle with an intermittent torque demand, the losses are also generated intermittently. In Figure 1.15, the torque pulses and the losses are assumed to follow the same waveform. The figure shows the limits of the steady-state, above-ambient temperature of the winding as Θ_{min} and Θ_{pk}.

If the shapes of the curves of winding temperature rise and fall over the pulse times t_p and t_s are assumed to be exponential, and to have the same time constant τ_W, we may write

$$\Theta_{pk} - \Theta_{min} = (R_{th}P_{loss(pk)} - \Theta_{min})(1 - e^{-t_p/\tau_W})$$

and

$$\Theta_{pk} - \Theta_{min} = \Theta_{pk}(1 - e^{-t_s/\tau_W})$$

Combining the last two equations and writing $t_p + t_s$ as t' gives the peak rise above ambient of the winding temperature as

$$\Theta_{pk} = R_{th}P_{loss(pk)} \frac{1 - e^{-t_p/\tau_w}}{1 - e^{-t'/\tau_w}}$$

The $i^2 R$ loss arises in the winding, which has a relatively low thermal capacity. The winding temperature rises faster than the rotor iron temperature, and also falls faster during the time t_s. The thermal time constant for the winding is therefore

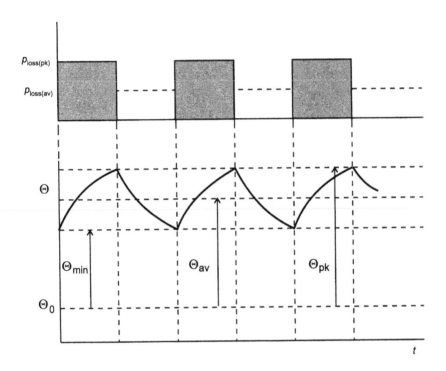

Figure 1.15
Steady-state temperature of the rotor winding

lower than for the rotor as a whole. The above expression can be used to predict the limiting conditions for the ripple at an assumed value of τ_w. If the average loss is kept at a constant level, the ripple on the winding temperature becomes more pronounced as t' is increased and/or as t_p is reduced. As a rule of thumb, the ripple in the steady-state temperature can normally be assumed to be within a band of $\pm 10°C$ when

$$\tau_{th} > 50t'$$

where $\tau_{th} \geq 25$ minutes. The thermal time constant of the motor used in the example of the sliding doors is given as 25 minutes, or 1500 seconds, and the period of the duty cycle in Figure 1.13 is 16 seconds. We can conclude that the winding temperature may be designed to reach 140°C simply as a result of the average i^2R loss, assuming there is no significant speed-sensitive loss.

Servomotor ratings

For both brushed and brushless servomotors, extra losses are generated if an application demands rapid changes in the speed of the motor and load. When rating the motor it is important to add the extra losses to those for the periods of steady speed, especially if the transient periods form a significant part of the duty cycle. We will look at such cases for the brushless servomotor in Chapter 5.

1.5 The brushed servomotor

So far we have studied the permanent magnet, brushed DC motor, mostly without reference to its role as a servomotor. In the example of the sliding door operator, the speed of the doors was determined by the balance between the motor output and the frictional forces developed in the door slide mechanism. No other control of the speed, or rate of change of speed, was required and the system can be described as open loop in the sense that speed control is achieved without the need for information feedback from the load to the motor.

For servo applications, precise control of load speed may be required at various stages of an operation and the servomotor must be capable of responding to calls for high transient torques. Two typical brushed servomotors are shown in Figure 1.16. The most striking difference between these and the normal DC motor is in the long and narrow shape, which gives the rotor a relatively low moment of inertia, increasing the output torque available for acceleration of the load itself.

The stators of the motors illustrated carry four permanent magnets made from a highly coercive ferrite material designed to withstand high demagnetizing fields. Also on the stator are four brushes which form the main point of motor maintenance. Depending on the motor duty, inspection is recommended up to eight times during the life of the brushes. The speed of a servomotor must be controllable at all times.

The speed is measured using the signal from a tachometer mounted on the motor shaft in the rear housing. The tacho has its own permanent magnetic field and brushes, and is a precision instrument which must be maintained in the same way as the motor itself.

Figure 1.16
Permanent magnet, brushed servomotors

The thermal characteristics of a typical DC servomotor are shown in Figure 1.17. The motor speed axis is marked in krpm, or revolutions per minute $\times 10^{-3}$. The curves are drawn for a winding temperature rise of 110°C. There are two continuous duty characteristics, one with and one without forced cooling. Both assume that the motor has a pure DC, unity form factor supply and derating may be needed if this is not the case.

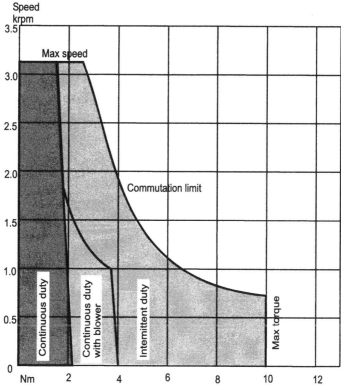

Figure 1.17
Thermal characteristics for a brushed servomotor

Example 1.1

Continuous operation is required at a speed of 1000 rpm. What is the maximum, average torque indicated by the Soac curve if cooling is unforced and the form factor of the current supplied by the electronic drive is 1.1? $K_T = 0.43$ Nm/A.

Figure 1.18 shows the type of current waveform provided by the electronic drive. The ripple is produced by the action of electronic switches as they operate to control the average value of current.

The maximum torque at 1000 rpm is found from Figure 1.17 to be 2 Nm. The maximum rms current which may be supplied to the motor at 1000 rpm is therefore

$$I_{rms} = T/K_T = 2/0.43 = 4.65 \text{ A}$$

The usable average current is

$$I_{av} = I_{rms}/\text{Form factor} = 4.65/1.1 = 4.23 \text{ A}$$

and so the maximum average torque is

$$T_{max} = 0.43 \times 4.23 = 1.82 \text{ Nm}$$

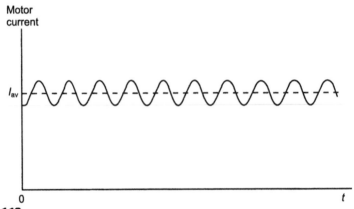

Figure 1.18
Current supply for Example 1.1

Example 1.2

A torque of 8 Nm at 500 rpm is required once every 9 seconds. What pure DC current would be required and what would be the maximum intermittent operation time? The maximum ambient temperature is 30°C and $\tau_{th} = 50$ minutes.

Figure 1.19 shows the current in this case is assumed to be ripple free, and to consist of a series of rectangular pulses of length t_p. The maximum, continuous DC at 500 rpm is

$$I = T/K_T = 2.05/0.43 = 4.77$$

At the torque of 8 Nm, the motor current is

$$I_p = 8/0.43 = 18.6 \text{ A}$$

For an average winding temperature rise of less than 110°C, the rms value of the intermittent pure DC should be no more than the maximum continuous value, and so

$$(t_{\text{pmax}} \times 18.6^2/9)^{0.5} = 4.77$$

giving

$$t_{\text{pmax}} = \mathbf{0.6 \ s}$$

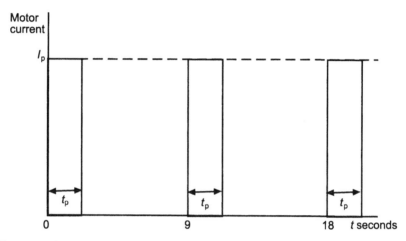

Figure 1.19
Current supply for Example 1.2

The 9 second period of the duty cycle is less than $\tau_{\text{th}}/50$ and we can assume that any steady-state temperature ripple will be less than $\pm 10°C$. The peak winding temperature is therefore less than $30 + (110 + 10)$, or $150°C$. Note that t_{pmax} would be less than 0.6 s if the current of 18.6 A is supplied as an average value by a source of impure DC, e.g. the electronic drive in example (1).

CHAPTER 2

THE BRUSHLESS MACHINE

2.1 Introduction

The design and characteristics of the brushed type of permanent magnet DC motor were reviewed in Chapter 1. It was seen that the action of the brushes and commutator ensures that the current flow through the rotor winding is always in the same direction relative to that of the permanent magnetic field. It was also found that the current in each turn of the winding has a rectangular AC waveform, alternating in direction as the winding rotates. One of the roles of the brushes and commutator is therefore to act as an *inverter*, converting DC from the power supply to rectangular AC for use around the turns of the winding. This is the key to the understanding of the brushless motor, where the brushes and commutator are replaced by an electronic inverter. The inverter separates the remaining electromechanical device from the DC supply and provides it with the required AC.

The layout of the normal brushed DC motor is fixed by the very use of brushes to effect reversals in the direction of the motor current. It is the rotor of the brushed motor which must carry the load current, and only the stator is available for the magnets. This construction has some disadvantages, not only in the commutation of heavy load currents but also in the generation of losses in the part of the machine most difficult to cool. In the brushless servomotor the load conductors can be placed on the stator and the permanent magnets, which need no external connections, on the rotor. The i^2R losses are more easily removed from this stator than is possible from the rotor of the brushed machine.

Electronic inverters and the method of supplying the current to the brushless motor are described in Chapter 3. This chapter is concerned with the brushless machine itself. The operation of the motor is dealt with from first principles using the basic electromagnetic fundamentals introduced in Chapter 1. The important features of the magnetic circuit of the motor are covered later on in the chapter.

2.2 Structure and operation

Figure 2.3 shows a complete brushless servomotor and Figure 2.1 shows its main components. The load conductors are wound on a stator core which is separated from the finned, aluminium case by an electrically insulating sleeve. Figure 2.1 shows two rotors of different designs.

Figure 2.1
Brushless motor component

Stator construction

The stator core is made of silicon steel, with slots for the current-carrying conductors. The core is laminated in order to minimize eddy currents. The laminations, one of which is shown in Figure 2.1, are punched from 0.3–0.5 mm steel sheet. The slots are seen to take up a relatively large part of the total area of the lamination, and have the effect of disrupting the uniformity of the path of the flux. The effect is reduced if the slots are skewed relative to the stator axis, as can be seen in the figure. The skewing presents an overlapping slot pattern and a less irregular path to the flux which enters and leaves the core radially. The windings can sometimes produce audible noise as they vibrate under the operating forces of the motor. An alternative construction is possible where the stator is rigidly bound by a resin mold, reducing its freedom to vibrate.

The permanent magnet rotor

The rotor hub carries the permanent magnets, and is pressed into position on the motor shaft. The hub can be machined from solid, low carbon steel or assembled from laminations punched from the centre of the steel sheet used for the stator laminations. The rotors shown in Figure 2.1 are of four-pole design. One has magnets of a cylindrical shape, and the other has magnets with non-circular surfaces. For a two-pole rotor with cylindrical magnets, the ideal flux density around the pole circumference would vary as a single-cycle rectangular wave, as shown in Figure 2.2. In practice some irregularity remains in the magnetic circuit even when the slots are skewed, and the dotted line along the top of the flux wave shows the effect of the irregularity on the waveform of the flux density in the air gap.

Magnets with a high flux density are used to maximize the torque/rotor volume ratio of the brushless servomotor. These must be very firmly fixed to the hub, and this has been one of

the main problems of manufacture. As well as the tensile radial
force on the magnets at high rotor speed, there are also shear
forces which must be resisted during abrupt acceleration and
deceleration. The magnets are bonded to the hub using
adhesives with mechanical and thermal expansion coefficients

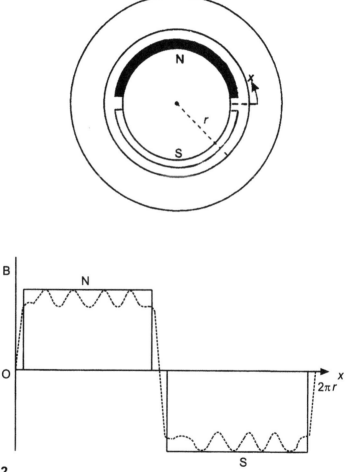

Figure 2.2
Flux density in the air gap around a two-pole cylindrical rotor

close to those of the magnet, and other devices are added to
ensure that the magnets do not destroy themselves and the

motor by parting company from the hub. Figure 2.3 shows one of the most common, where the magnets are bound by fibreglass tape. Figure 2.1 shows the rotors before the tape and end rings have been fitted. The extra manufacturing cost involved in bonding and binding the magnets to the hub is a small drawback to the permanent magnet rotor.

Figure 2.3
Brushless motor and rotor

Motor operation

The typical stator lamination shown in Figure 2.1 has deep slots for the stator conductors, and the conductors are therefore not in the air gap between the rotor and stator surfaces. In Chapter 1, a simple explanation of the operation of the brushed motor was given without reference to the slotted nature of the rotor. When the torque mechanism was described, the rotor conductors were assumed to cut the air gap flux around the pole faces. Before using the same method with the brushless motor, we should first look at whether or not the assumption is valid.

Flux cutting and flux linkage

In Chapter 1, expressions for torque and emf were found using the flux-cutting approach. With this method, the starting point in the derivation of emf is the expression

$$e = Blv$$

Flux of density B is assumed to be cut by a conductor of length l moving at velocity v, or to pass across one that is stationary. Figure 2.4 shows the conductors of a brushless motor winding lying in a stator slot and almost surrounded by the stator iron. The objection to the flux-cutting method which is sometimes raised is that most of the flux does not cross the conductor but instead passes around the slot, through the iron.

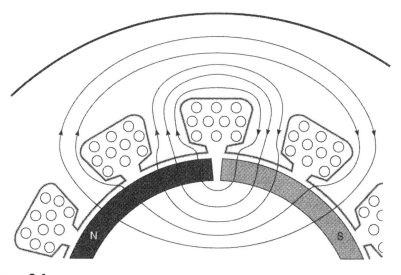

Figure 2.4
Flux passing around the stator slots

The flux-cutting method overcomes this problem by assuming that all the conductors are effectively on the inner surface of the stator, in the air gap flux at a radius of action equal to half the diameter of the stator bore. With these assumptions, it is found that the method gives the correct basic results for motor torque and back emf. The conceptual difficulty with the flux-cutting approach is avoided in the flux-linkage method. Faraday's law is used to express the emf induced across one metre of a conductor as

$$E = \frac{d\lambda}{dt}$$

where λ is the quantity of flux which links the conductor completely at time t. Both methods lead to identical results for the basic analysis of motor torque and back emf, and so we will continue to use the flux-cutting approach.

Torque

Figure 2.5 shows four positions of a rotor, relative to the stator of a simple two-pole brushless motor. The associated directions of the stator current are shown by using the circled cross and dot convention for current flowing into and out of the paper. A stator with a single winding has been chosen for the purpose of explaining the operating principle. The stator of a brushless servomotor normally has three windings, but the principle is the same.

In Figure 2.5, the current direction in the winding is reversed (using external means) at positions 2 and 4, when the pole centres pass from one side of the winding to the other. Consequently, at rotor angle θ the average of the forces BIl on the current-carrying conductors is always in the same direction, except at 90° and 270° when the overall force is zero. The resulting torque cannot cause rotation of the stator and so the equal and opposite reaction at the rotor provides output torque to the load. The torque is produced in the same way as in the brushed machine and we can express the average output as

$$T = K_T I$$

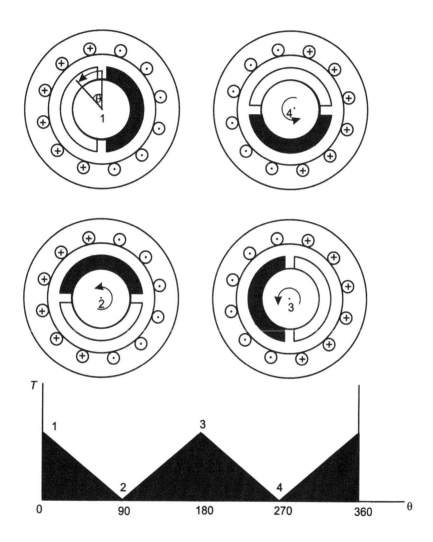

Figure 2.5
Torque from a simple brushless motor

The peak torque at positions 1 and 3 in Figures 2.5 and 2.6 is equivalent to the steady output torque of the two-pole brushed motor already described in Figure 1.3. The average torque constant of this particular brushless machine is therefore only half that of the brushed motor. The ripple in the output torque can be described as 100% and would be unacceptable in most servomotor applications. The problem arises when the poles are halfway between one side of the winding and the other, where the net force is zero. We will see shortly how the use of three windings eliminates such nulls to produce a theoretically constant output.

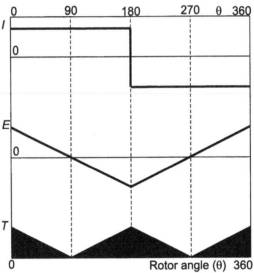

Figure 2.6
Torque and back emf for a single-winding brushless motor

Back emf

Rotating flux sweeps across the stator winding to produce a back emf with an average half-cycle value of

$$E = K_E \omega$$

where K_E is half the value of the voltage constant of the brushed machine of Figure 1.3. The back emf alternates in direction as the poles of the magnet change position, as shown in Figure 2.6. It is important to note that the back emf at the input terminals of the brushless motor alternates in direction, as does the direct current input. The motor has the same construction as an AC synchronous motor, which normally has a sinusoidal rather than rectangular current waveform.

Although the single-phase brushless machine works correctly as a motor, its output torque is 'lumpy' and would be unsuitable for most industrial servomotor applications. The main uses occur at the low power end of the scale where the brushless motor is manufactured with a single winding in very large numbers, for example as fan motors for the cooling of electronic equipment. These are normally exterior-rotor motors, where the fan is mounted on a hollow, cylindrical permanent magnet which rotates around a laminated, cylindrical stator with slots for the winding.

2.3 The three-winding brushless motor

Most industrial brushless servomotors have three windings, which are normally referred to as phase windings. There are two main types. One is known as the *squarewave* motor, the name being derived from the (theoretically) rectangular waveform of the current supplied to its windings. The other is supplied with sinusoidal AC and is known as the *sinewave* motor. Both types are physically very similar to the three-phase AC synchronous motor.

The squarewave motor

The windings of the ideal squarewave motor would be supplied with currents in the form of perfectly rectangular pulses, and the flux density in the air gap would be constant around the pole faces. The squarewave version of the small four-pole motor in Figure 2.3 would have the cylindrical magnet rotor shown in Figure 2.1. Figure 2.7 shows a simple layout for a two-pole machine where each of the three windings, a, b, c, is divided into two coils connected in series; for example, coils b_1 and b_2 are connected in series to form winding b. The start and finish of, for example, coil b_1 are marked b_1 and b_1'. The two coils of each winding have an equal number of turns and are mechanically spaced apart by 30° around the stator.

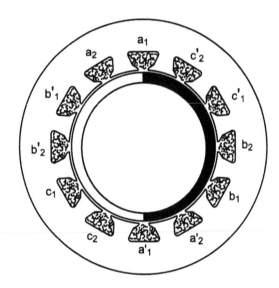

Figure 2.7
Two-pole, three-phase motor with two slots per phase, per pole

The effect of distributing each winding into more than one slot is to extend the arc over which the winding is influenced by each pole as the rotor turns. This means that the number of slots should be specified in relation to the number of poles as well as to the number of windings. The stator in Figure 2.7 is symmetrical, with three windings, 12 slots and six coils each with an equal number of turns. As a result, each phase will provide the same magnitude of torque and back emf.

Torque production per phase

Figure 2.8 shows how the a-phase torque is produced in the squarewave motor when the current has the ideal rectangular waveform shown. The method of supplying the current and its commutation between the phases is described in Chapter 3. The flux density waveform around the pole faces has not been shown in a rectangular form. Changes in flux direction are less abrupt due to the skewing of the stator slots, and the flux waveform is shown in the diagram with ramp leading and trailing edges as a first approximation. In practice the corners are rounded due to fringing effects near the edges of the poles. Rotation is anticlockwise and the coincidence of the pole divisions with the first coil has been chosen as the starting point. The flux direction is drawn for the N-pole, and the current direction for the upper coil sides in the diagram.

As the rotor moves from the $\theta = 0°$ position, N-pole flux starts to cross the upper side of the first coil, and when $\theta = 30°$ the second coil comes under the same influence. The lower sides a' of the coils are similarly affected by the S-pole flux. As the rotor turns through 180°, the flat topped section of the flux wave moves across the full winding over a window of 120°. This is the period when the current must be fed in from an electronically controlled supply. Positive torque is produced as the current flows through the winding. The cycle is completed as θ changes from 180° to 360°, again producing positive torque.

Magnitudes of back emf and torque per phase

The 'ac' nature of the back emf is evident in Figure 2.8. When the flat topped part of the flux wave sweeps across the coils of the 'a' phase, the voltage generated across one side of one turn of either coil is

$$e' = Blr\omega$$

where the speed of rotation is ω rad s^{-1}, and l is the length of the coil side (into the paper). The voltage generated around a complete turn is

$$2e' = 2Blr\omega$$

If the winding has N_{ph} turns distributed between the two coils, the total back emf generated around the two series-connected coils is

$$e_a = 2N_{ph}Blr\omega$$

The torque produced by one side of one turn of either coil is

$$t' = Bli_a r$$

and the total a-phase torque is

$$t_a = 2N_{ph}Bli_a r$$

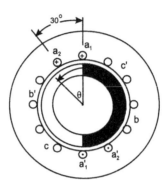

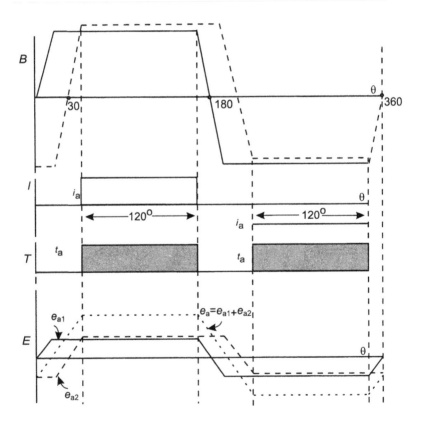

Figure 2.8
Torque and back emf for the 'a' phase

The three windings are symmetrically distributed around the stator, as are the magnetic poles around the rotor, and so

$$e_a = e_b = e_c$$

and

$$t_a = t_b = t_c$$

Before combining these quantities to give the output torque and the back emf at the input terminals, we should first look at how the motor windings are connected together.

Wye (Y) and delta (Δ) connections

Figure 2.9(a) shows the Y or star connection, where the windings are joined to form a star point. The figure also shows the motor currents which flow from an electronically controlled source. Each winding of the star is in series with its supply line, and the same current flows in the line and the winding. One full cycle of each phase current must occur for every 360° of rotor movement and so i_b and i_c are displaced from i_a by $\theta = 120°$ and 240°. Note that the sum of the three currents at the star point is zero for all values of θ. Note also that the emf across a pair of motor terminals is the difference (for the chosen reference directions) of the emfs across the respective phase windings.

Figure 2.9(b) shows the Δ connection, where the emf across the windings appears across the motor terminals. The line currents are the same as before but differ here from the phase currents. The difference between any two phase currents equals the line current flowing to the common point of the two windings. The line-to-line voltages are no longer trapezoidal, and the phase emfs do not sum to zero. Circulating currents are likely around the closed delta path, with the possibility of motor overheating due to the extra i^2R losses. The Δ connected stator is therefore less useful and most squarewave motors are made with the Y connection.

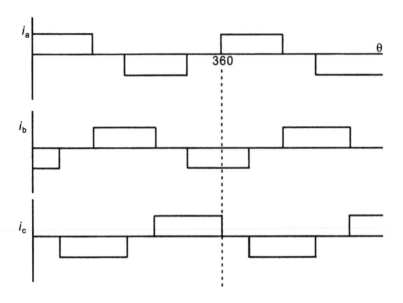

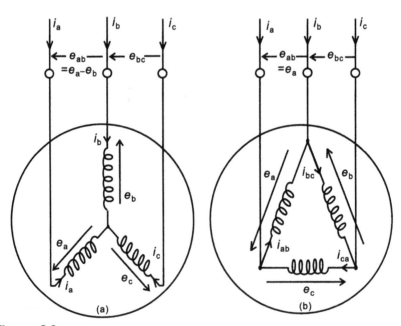

Figure 2.9
Effect of motor connections on phase currents and voltages

Three-phase torque and back emf

Figure 2.10 shows the patterns of ideal torque and emf for each of the three windings of a Y-connected squarewave motor with the winding and pole layout described in Figure 2.8. The squarewave motor is often referred to as the 'trapezoidal' motor in view of the trapezoidal shape of the back emf. The emf across the a–b input terminals in Figure 2.9(a) is

$$e_{ab} = e_a - e_b$$

and so the peak emf in Figure 2.10 is

$$e_{ab} = 2e_a$$

or

$$e_{ab} = 4N_{ph}Blr\omega$$

The back emf across a pair of machine terminals is

$$e_{ab} = e_{bc} = e_{ca} = E = K_E\omega$$

where $K_E = 4N_{ph}Blr$, the voltage constant of the motor.

Looking now at the patterns of torque produced by the motor, we see that each phase works for 240° and rests for the remaining 120° of each turn of the rotor. However, the combined effort of the three phases does produce the extremely important feature of a theoretically smooth output torque. Only two phases produce torque at any one time and so the motor torque is

$$T = 2t_a$$

or

$$T = 4N_{ph}Bllr$$

where I is the line current input to the motor. The torque can be written in the familiar form

$$T = K_T I$$

where $K_T = 4N_{ph}Blr$, the torque constant of the motor.

Comparison between the emf and torque expressions confirms that the voltage and torque constants are equal for the squarewave motor. As in the case of the brushed motor, the numerical equivalence exists only when the constants are expressed in SI units.

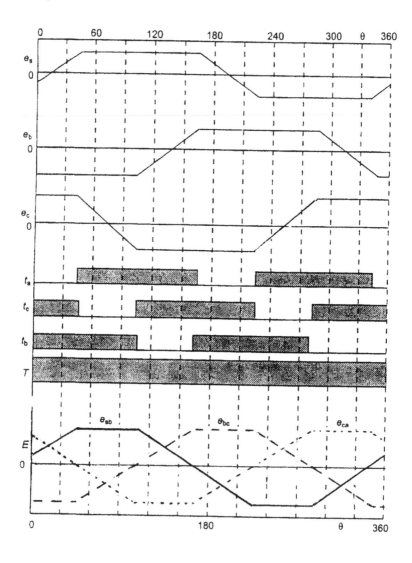

Figure 2.10
Emf and torque for a Y-connected squarewave motor

Practical emf and torque waveforms

The smoothness of the rotor output torque is affected by fringing effects which leave less than 120° of the flux wave in a flat-topped form. This is in addition to the effect of the ripple in the flat top caused by stator slotting. Further irregularities in the output torque result from stator current waveforms which are never perfectly rectangular in practice.

The sinewave motor

The ideal squarewave motor has rectangular waveforms of flux density and input current, and has windings concentrated in coils in the stator slots. The ideal sinewave motor has sinusoidal flux and current waveforms and a sinusoidal distribution of its windings.

Sinusoidal AC input current

In common with squarewave motors, most sinewave machines are made with three phases. Figure 2.11(a) shows the three line currents which are supplied to the motor from an electronic inverter.

Sinusoidal flux density in the air gap

There are a number of ways in which the magnetic circuit can be designed to produce a near sinusoidal flux density waveform. A good sinewave can be achieved by tapering the magnets towards the edges as shown in Figure 2.11(b). The taper of the profile is exaggerated in the diagram. Formation of the waveform is assisted by fringing effects which are encouraged by the use of a relatively small pole arc. Figure 2.1 shows a four-pole, sinusoidal rotor where the tapered magnets are mounted on a square section hub.

Sinusoidal winding distribution

The ideal, fully distributed layout of stator conductors for one phase of a two-pole, sinusoidal motor is represented in Figure

2.11(c). In practice an irregularity must be present in the distribution due to the bunching of conductors in slots.

The three-phase sinewave motor closely resembles the three-phase AC synchronous motor and its characteristics can be found through the phasor diagram method. However, its ideal torque and back emf can still be found by the method we have used for the squarewave machine. Figure 2.12 shows one phase of a two-pole sinewave motor with ideal flux, current and winding distributions. There are N_s conductors on each side of the winding of N_s turns. The reference has been chosen at the moment when the N–S pole axis of the rotor lies horizontally in the diagram, when $\theta = 0$ and the input current is zero. We assume that the current is controlled (externally) in such a way that it varies sinusoidally with rotor angle θ. Note that the current magnitude varies with θ and not with stator angle ϕ. The diagram is drawn for a moment in time when $\theta = 90°$, and the conductor current is therefore at its maximum of I_M.

Back emf

When $\theta = 90°$, the emf across a conductor of length l at stator angle ϕ is

$$e_1 = Blv$$

or

$$e_1 = B_M \sin \phi \, lr\omega$$

The combined emf across the conductors within $d\phi$ is

$$e_{d\phi} = \frac{N_s}{2} \sin \phi \, d\phi \, B_M \sin \phi \, lr\omega$$

or

$$e_{d\phi} = \frac{N_s}{2} B_M lr\omega \sin^2 \phi \, d\phi$$

Integrating this expression over $\phi = 0$ to π gives the total back emf across N_S conductors as

$$E_M = \frac{\pi}{4} N_S B_M lr\omega$$

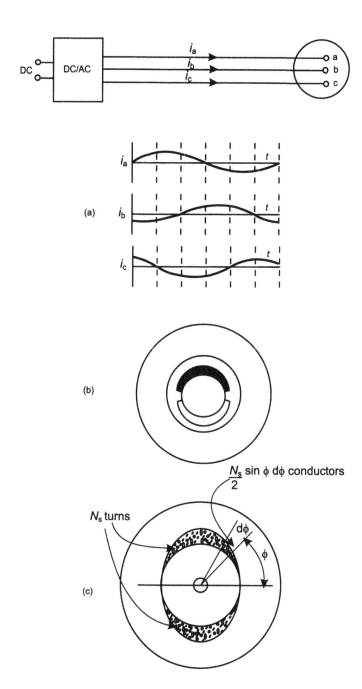

Figure 2.11
Main features of the sinewave brushless motor

E_M is the peak back emf, generated at the moment when flux cutting is at a maximum. The peak emf for the full winding, which has $2N_S$ conductors, is $2E_M$. The conductors are swept by a sinusoidally distributed flux and the variation of back emf with time must also be sinusoidal. The rms value of the emf across the full winding is therefore

$$E_{ph} = \frac{2E_M}{\sqrt{2}}$$

or

$$E_{ph} = \frac{\pi}{2\sqrt{2}} N_S B_M lr\omega$$

For a Y-connected sinewave motor, the back emfs across the three individual windings form a balanced set of three-phase voltages. The rms emf which appears across the motor input terminals and supply lines will therefore be $\sqrt{3}E_{ph}$, or

$$E = \frac{\pi\sqrt{3}}{2\sqrt{2}} N_S B_M lr\omega = K_E\omega$$

Torque

When $\theta = 90°$, the force on a conductor at stator angle ϕ is

$$f = BlI_M$$

or

$$f = B_M \sin \phi lI_M$$

The combined force on the conductors within $d\phi$ is

$$f_{d\phi} = \frac{N_s}{2} \sin \phi \, d\phi \, B_M \sin \phi \, lI_M = \frac{N_s}{2} B_M I_M l \sin^2 \phi \, d\phi$$

Integrating over $\phi = 0$ to 2π, the total force on the full winding of $2N_S$ conductors at the moment when $\theta = 90°$ is found to be

$$F_M = \frac{\pi}{2} N_S B_M I_M l$$

Figure 2.12 is drawn for the moment when the force on the winding is at a maximum. The rotor experiences an equal and opposite reaction to give a peak output torque of

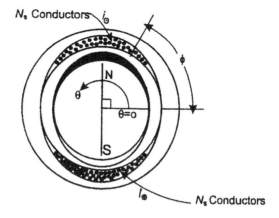

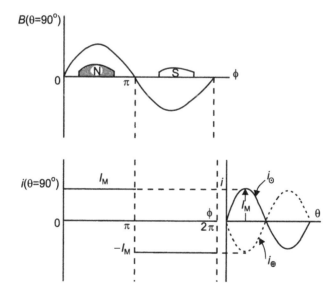

Figure 2.12
Flux density and phase current in the 2-pole sinewave motor

$$T_M = \frac{\pi}{2} N_S B_M I_M lr$$

T_M is the torque when the rotor pole axis lies at an angle of 90°
to the chosen reference position. When the rotor axis lies at θ to
the reference, the stator current is $I_M \sin\theta$ and the flux density
effective over the full winding is $B_M \sin\theta$. The torque becomes

$$T_\theta = T_M \sin^2\theta$$

T_θ is always positive and varies sinusoidally with rotor position, as shown in Figure 2.13.

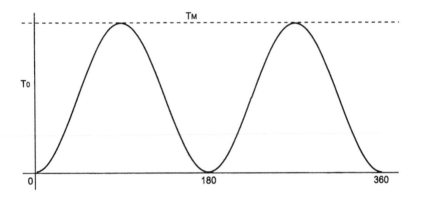

Figure 2.13
Torque from one phase of a sinewave motor

Three-phase torque and emf

The three-phase sinewave motor has three of the single-phase windings shown in Figure 2.12. The axis of each winding is 120° from the other two, around the stator. The waveforms of torque against θ produced by the three phases will therefore be separated by the same angle. The contribution of each phase to the overall torque is shown in Figure 2.14. The torque produced by each phase varies sinusoidally around its average value. When the three waveforms are added together, the sinusoidal components cancel out and we are left with the sum of the averages as the constant value of the torque on the rotor. The three-phase torque for the two-pole motor is therefore given by

$$T = \frac{3}{2}T_M$$

or

$$T = \frac{3\pi}{2\sqrt{2}}N_S B_M l r I_{rms}$$

In the above analysis, the torque of the ideal two-pole, three-phase sinewave motor has been found to have the same value for any particular position of the rotor poles relative to the stator windings. It has also been assumed that the waveform of winding current against time is synchronized with the sine of the rotor angle. Reference [1] gives a rigorous treatment for rotating poles by multiplying the total rotating ampere-conductor distribution of the stator by the rotating flux distribution of a rotor with p number of pole pairs. The torque is found to be dependent on the cosine of the angle by which the rotor lags behind the rotating field produced by the stator, but independent in the ideal case of the number of pole pairs. The effects of the non-ideal features of the practical windings are also covered in detail.

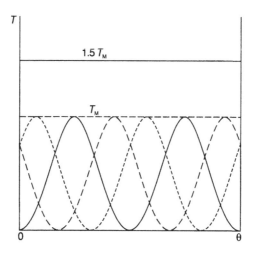

Figure 2.14
Combined three-phase torque from the sinewave motor

K_T and K_E

The last expression above gives the torque constant for the ideal sinewave motor as

$$K_T = \frac{3\pi}{2\sqrt{2}} N_S B_M lr$$

The voltage constant has already been seen to be

$$K_E = \frac{\pi\sqrt{3}}{2\sqrt{2}} N_S B_M lr$$

Comparing the expressions for K_T and K_E shows that the constants are not numerically equal for the sinewave motor. The relationship between the numerical values of the constants is given by

$$K_T = \sqrt{3}K_E$$

This form of the relationship is valid when K_E is expressed as Vrms/rad s^{-1} and K_T is the total three-phase torque constant in Nm/A$_{rms}$. Other forms are possible when the units are given in terms of peak, or per-phase values [2].

Torques and ratings

The three Y-connected windings are manufactured in squarewave or sinewave form. Both forms of the winding have the same resistance between the motor terminals. This means that the resistance per phase is $0.5R_{LL}$ in both cases, where R_{LL} is the value quoted by the manufacturer for the resistance between any two terminals as seen from the supply lines. Both forms of the motor have the same physical size, thermal resistance and torque ratings. At any time, the squarewave motor carries current in only two of its phase windings. The maximum i^2R loss is therefore

$$2I_{sq}^2 \frac{R_{LL}}{2} = I_{sq}^2 R_{LL}$$

where I_{sq} is the maximum, continuous rms current which may be carried by the two conducting windings without overheating the motor. For equal i^2R losses in the sinewave and squarewave versions of the motor we may therefore write

$$3I_{sn}^2 \frac{R_{LL}}{2} = I_{sq}^2 R_{LL}$$

or

$$I_{sn} = \frac{\sqrt{2}}{\sqrt{3}} I_{sq}$$

where I_{sn} is the maximum, continuous rms current which may be carried by each of the three conducting windings of the sinewave motor. The motors are designed to have the same torque ratings and so

$$K_{T(sn)}I_{sn} = K_{T(sq)}I_{sq}$$

Combining the last two equations above gives

$$K_{T(sn)} = \frac{\sqrt{3}}{\sqrt{2}} K_{T(sq)}$$

where the torque constants refer to the total output torque produced by each form of the motor. Care is needed when the torque constant of the sinewave type is defined. The constant is usually given for the total three-phase torque, but is sometimes given per phase. Table 2.1 shows some figures for the two forms of the small brushless motor shown in Figure 2.3. The last row shows the maximum, continuous current which can be supplied without overheating any part of the motor, when the rotor is locked in a stationary position. The continuous stall torque is similarly defined.

Table 2.1 Stall torques and currents for a small brushless motor

Input current waveform		Sinewave	Squarewave
Resistance	$R_{LL}\Omega$	5.4	5.4
Continuous stall torque	T_s Nm	2.2	2.2
Torque constant	K_T Nm/A	0.75	0.61
Continuous stall current	I_s A	2.9 rms	3.6 DC

The effective resistance

In the trapezoidal form of the motor, the current flows through only two phases at any one time and so the line-to-line resistance is the resistance which is effective in generating the i^2R loss. In the sinusoidal case, the loss is generated in the three phases at all times, to give a total of

$$3 \times I^2 \frac{R_{LL}}{2} = I^2 \times \frac{3}{2} R_{LL}$$

The thermally effective resistance of the sinewave motor is therefore $1.5R_{LL}$.

2.4 Permanent magnets and fields

Up to this point our attention has been on understanding how squarewave and sinewave motors operate, and the 'permanent' field of the magnets has been taken for granted. In the first part of this section we will look at permanent magnets in general, and later at those used in brushless servomotors.

Magnetic fields

The idea that a magnetic field consists of flux Φ of density B was used in Chapter 1, when the production of torque and back emf was explained for the brushed DC motor. There are three other concepts which are used in the description of the magnetic fields of both brushed and brushless motors.

Magnetomotive force mmf

The force which pushes magnetic flux along its path is called the *magnetomotive force*, or *mmf*. The air-cored and iron-cored coils in Figure 2.15 have N turns and carry current I. The force driving the flux is expressed as

$$mmf = NI \; (A\text{-turn})$$

For example, if each coil has 10 turns and carries 100 A, the mmf will be 1000 A-turn in both cases.

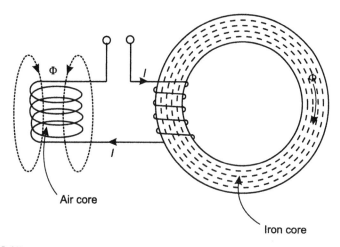

Figure 2.15
Mmf and flux in air- and iron-cored coils

Magnetic field intensity H

The shorter the path of the flux in whatever medium, the greater the amount of flux which can be established in that medium by a given mmf. The mmf per metre of the flux path is called the *magnetic field intensity*, expressed as A/m.

Permeability μ

As its name suggests, *permeability* tells how easy it is for the mmf to establish flux in a particular medium. The permeability of the medium is

$$\mu = \mu_o \mu_r$$

where μ_o is the permeability of a vacuum expressed in the unit of henry per metre and μ_r is the permeability of the medium relative to that of a vacuum.

The density of the flux in a particular case can be seen to be dependent on two factors. One is the intensity of the mmf around the flux path, and the other the permeability of the medium. The flux density is given by

$$B = \mu_o \mu_r H$$

In Figure 2.15, H will have about the same value in the two cases. However, μ_r for iron is of the order of several thousand and so the flux density will be much greater here than in the air-cored coil, where μ_r is close to unity.

The hysteresis B–H loops

The 'normal' characteristic which describes the properties of a permanent magnet is the B–H loop shown in Figure 2.16. The dotted line is the so-called 'intrinsic' loop. The normal curve shows the full cycle of magnetic states which can be induced in the magnet, starting at the origin with a previously unmagnetized sample. Figure 2.17(a) shows such a sample clamped between the ends of an iron core, so that the external magnetizing force NI A-turn can be applied. Note that the x-axis in Figure 2.16 is not scaled as H but as $\mu_0 H$, which is the flux density which would exist in the air between the ends of the iron core without the magnet in place. The y-axis gives the density which actually appears in the magnet when in place. Let us now go round the normal B–H loop of Figure 2.16, starting at the origin. Assume that the iron core has a very high permeability. This allows the full mmf to appear across the ends of the specimen, without loss along the iron path.

0–A Mmf NI and field intensity H (between the ends of the iron core) are increased from zero until the flux density B in the magnet reaches a maximum at A.

A–B_r Current I is switched off at A upon which the flux falls to a residual level of density B_r, and not to zero.

B_r–C The externally-applied mmf is again increased from zero, but this time in the negative direction. The flux density within the magnet falls to zero at C.

C–D The negative current in the coil is further increased and flux density rises from zero towards its negative peak.

D–A The external mmf is returned to positive values. The D–A return path is usually a mirror image of the A–D route.

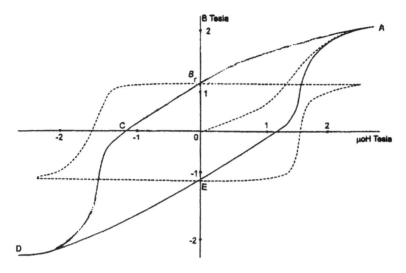

Figure 2.16
Hysteresis loops for a brushless servomotor, permanent magnet

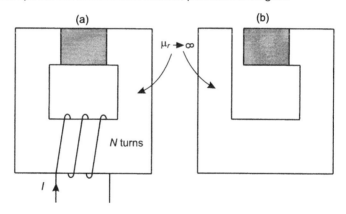

Figure 2.17
The magnet as part of two magnetic circuits

Suppose now that the coil is removed, so that B falls to B_r, and then an air gap introduced into the magnetic circuit as shown in Figure 2.17(b). The value of B around the magnetic circuit will fall below B_r due to the low permeability of the gap. The circuit of a brushless motor consists of the magnets in series with paths through the rotor and stator iron, and the gaps between the pole faces and the stator. The working density must therefore be less than B_r.

The operating quadrant of the motor is shown in Figure 2.18. Note that the straight line from B_r passes through C, and that the knee of the curve is below the horizontal axis. The straight section is the most important feature of the so-called 'hard' materials used in the fabrication of magnets for high performance, brushless servomotors. The term refers to a magnetic rather than a physical toughness. The permanent field of the hard magnet will not be damaged by stray fields provided the flux density in the magnet is not forced to a point on the knee of the curve. The intrinsic curve is used to interpret this behaviour.

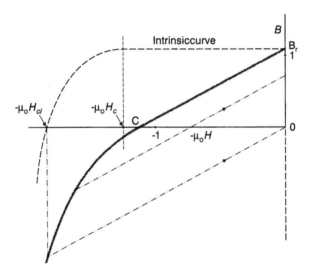

Figure 2.18
The operating quadrant of the hard magnet

The intrinsic curve

The vertical axis of the intrinsic curve of Figure 2.18 gives the flux density which is potentially still available in the magnet after the externally applied field of intensity H is removed. Certain points can be defined:

Remanence B_r   The maximum intrinsic flux density when the permeability of the external flux path is infinite.

Coercivity H_c   The maximum external field intensity which does not cause demagnetization which is intrinsically irrecoverable. The value for $\mu_o H_c$ is given at the lower end of the straight-line section.

Intrinsic coercivity H_{ci}   The applied field intensity which completely demagnetizes the magnet.

For the magnets of the brushless motor, the most important feature of Figure 2.18 is that the intrinsic flux density remains at B_r following the application and subsequent removal of reverse fields with intensities up to H_c, but falls below B_r when the reverse field has an intensity above H_c.

Permeability of the hard magnet

The vertical axis of the *B–H* loop of Figure 2.18 shows the flux density in tesla which is set up in the magnet by the application of an mmf of magnetic intensity H across its ends. The horizontal axis is also scaled in tesla, and gives the flux density $\mu_o H$ which would exist in the air space occupied by the magnet. For hard magnetic materials the slope of the straight-line working section is $B/\mu_o H \approx 1$, and so for the magnet itself we have

$$B \cong \mu_o H$$

The surprising conclusion is that hard magnets have a low permeability, close to that of air. They are said to have a low recoil permeability. This term is not particularly good as it suggests that the magnet has low rather than high intrinsic ability to recover its flux levels.

Armature reaction

In Chapter 1 we reviewed the permanent-magnet brushed DC motor, where the current-carrying conductors are wound on the rotor. The rotor of the brushed motor is often called the armature of the machine. The flow of current around the armature winding sets up a magnetic field which combines with the field produced by the stator magnets or field windings of the brushed motor. The resulting distortion of the field in the air gap is described as being due to *armature reaction*. The same term is applied to the brushless motor, even though the source of the effect is the stator and not the rotor.

The mmf developed by the stator winding of the brushless motor may be quite high, especially at full load current. Flux circulates around the conductors, some staying in the stator and some crossing the air gap to pass through the magnet. However, we have seen that the magnet has much the same permeability as the air gap. The result is that the relatively small amount of stator-induced flux which enters the magnet has little effect on the average operating level of the flux density in the air gap, although it does cause some distortion in the flux distribution.

Overload currents

We shall see in the next chapter how the current is supplied to the brushless motor. The supply units use power electronics to control the flow of current, normally with a very high reliability. However, faults can never be entirely ruled out. The worst effect for the permanent magnet machine is likely to be from the magnetic fields set up by the flow of high overload current through the stator conductors. These fields are obviously stronger than those due to normal armature reaction and the main concern must be to avoid the risk of any permanent effects on the strength of the magnets. Fortunately the hard magnet has some protection of its own through low permeability and high coercivity. These features are exploited when the cost of the magnet is minimized by reducing the radial length (i.e. in the magnetized direction) to the minimum necessary.

Permanent magnet materials

Two materials have become well established in the manufacture of hard permanent magnets for industrial brushless servomotors.

Samarium cobalt (Sm–Co)

Samarium and cobalt are both rare-earth elements. There are only a few sources of samarium in the world which can supply the quantities needed and so the cost is invariably high. Even so, the Sm–Co magnet is widely used in brushless servomotors. This is mainly because it has superior technical characteristics when compared to ferrite, and is particularly good when compared to metal alloys such as Alnico.

Neodymium iron boron (Nd–Fe–B)

The 'Nib' magnet uses materials which are less expensive than samarium and cobalt. The lower cost of the magnet is, however, its only advantage, as it has no better technical characteristics for motors than the Sm–Co type. One of the most troublesome problems of the Nib magnet is its susceptibility to corrosion and although it has a better second quadrant B–H characteristic than Sm–Co, the advantage is lost at the high end of the operating temperature range.

Temperature effects

The Sm–Co magnet has a better high temperature coercivity and a better temperature coefficient of remanence than the Nib type. In Figure 2.18, the knee of the B–H characteristic lies below the H-axis. As temperature rises, the knee moves up the curve with the danger of intrusion into the operating quadrant. Figure 2.19 shows the relative effects of the knee movements for the two materials in question. The knees below the H-axis move in a way which brings both characteristics closer to the origin, but the movement of the Sm–Co line is small in comparison to that of Ne–Fe–B. The knee of the Sm–Co characteristic does not move up enough to affect the linearity in the operating quadrant, but the movement of Ne–Fe–B is greater and linearity is not maintained. This means that the effects, at typical motor temperatures, of demagnetizing fields with intensities up to the 'hot' value of coercivity will not be permanent for the Sm–Co magnet but may be so for Ne–Fe–B. It should be remembered, however, that even the Sm–Co magnet can still be demagnetized if the maximum allowable motor temperature is exceeded.

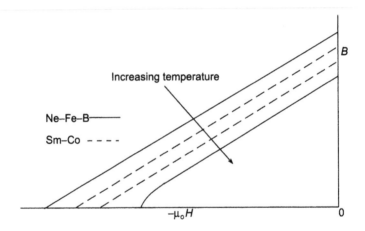

Figure 2.19
Demagnetization of Sm–Co and Ne–Fe–B magnets

2.5 Characteristics

The structure of the brushless machine gives it significant advantages in performance over the brushed type. Performance is enhanced because:

1. There is no brush and commutator transmission of current to the motor and therefore no mechanical commutation limit to the speed at which any particular torque can be supplied.

2. The i^2R loss arises in the stator rather than the rotor, allowing the surplus heat to pass more freely into the air surrounding the motor case. Any overheating which does occur is also easier to detect as the effects occur in the accessible component.

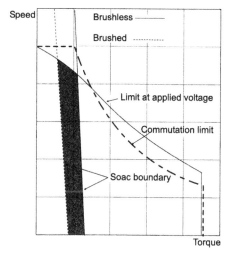

Figure 2.20
Thermal characteristics

There are several ways of comparing the thermal characteristics of brushed and brushless motors. Perhaps the fairest is to compare motors of about the same physical size, the same maximum torque and the same speed range when operated from the same voltage. Figure 2.20 shows such a comparison, the shaded area depicting the higher continuous torques available from the brushless motor. Apart from a small region close to maximum speed, the levels of continuous torque available from the brushless motor are of the order of 40% greater than from the brushed type. The intermittent torques available are higher for the brushless motor over most of the diagram, and are generated without the brush and commutator deterioration suffered by the brushed motor when working near to the commutation limit.

Specifications

The physical size of the most commonly used motors varies quite widely, with motor weights from as little as 1 kg to a substantial 50 kg. The maximum continuous power outputs vary from 50 W to 10 kW, and up to four times these figures intermittently. Three motors from the medium to small end of the range are shown in Figure 2.21. The smallest has a length of approximately 12 centimetres. The largest of the three can supply a continuous power demand in the region of 2.5 kW at a nominal speed of 3000 rpm, and its specification includes the details shown in Table 2.2. The motor is a four-pole machine with Sm–Co magnets and is manufactured in either squarewave or sinewave form. Here we should remember that such names refer to the ideal current waveforms. Motor specifications normally classify the motors according to the shape of the back emfs, and the table shows a choice between the *trapezoidal* and *sinusoidal* types.

Figure 2.21
Brushless servomotors

Trapezoidal

For the trapezoidal form of the motor in Table 2.2

$$T_S = K_T I_S = 0.84 \times 11.7 = 9.8 \text{ Nm}$$

The data given on the specification sheet does not normally include a value for the voltage constant in SI units. However, K_T and K_E should have the same numerical value for the trapezoidal motor, and this can be checked from the table. The back emf is

$$E = K_E \omega$$

or

$$88 = K_E \times 1000 \times 2\pi/60$$

giving

$$K_E = 0.84 \text{V}/\text{rad s}^{-1}$$

Table 2.2 Brushless motor constants

		Sinusoidal	*Trapezoidal*
Peak line-line emf	V/krpm	88.0	88.0
Continuous stall torque	T_S Nm	9.8	9.8
Continuous stall current	I_S A	9.6 rms	11.7
Torque constant	K_T Nm/A	1.02	0.84
Resistance	R_{LL} Ω	1.01	1.01
Max. cogging torque	Nm	0.24	0.32
Overall length	mm	315	315
Weight	kg	11.4	11.4

Sinusoidal

The continuous stall torque is

$$T_S = K_T I_S = 1.02 \times 9.6 = 9.8 \, \text{Nm}$$

We know that K_T and K_E do not have the same numerical value for the sinusoidal motor. For the motor in question, the rms line-to-line emf per 1000 rpm is

$$88/\sqrt{2} = K_E \times 1000 \, 2\pi/60$$

giving

$$K_E = 0.59 \text{ Vrms}/\text{rad s}^{-1}$$

For the sinusoidal form of the motor, we have the result that

$$K_T/K_E = 1.02/0.59 = 1.73$$

This agrees with the theoretical relationship; $K_T = \sqrt{3}K_E$.

Cogging torque

Table 2.2 includes a figure for cogging torque. A constant output of torque depends on the waveform of flux density in the air gap being either perfectly flat-topped, or perfectly sinusoidal. Stator slotting produces a non-uniform magnetic flux path which in turn affects the flux patterns and gives rise to a rippling or cogging of the torque. There are two main ways of dealing with the problem:

Fractional slotting   The motor in Figure 2.7 has 12 slots and two poles, making an integral slot/pole ratio. For any such integral ratio, all pole edges would line up with slots at the same time as the rotor turns. A reduction in the cogging effect can be made by forming a non-integral slot/pole ratio to produce the so-called fractional-slot winding. It is, however, difficult to make a trapezoidal motor in such a way, and still maintain the necessary length of the flat-topped section of the rotor flux wave. The technique is easier in the sinewave motor, where fractional slotting is combined with an uneven spacing of the magnets in order to reduce the cogging effect still further.

Skewing   The stator slots can be skewed relative to the rotor axis in order to reduce the irregularity of the magnetic circuit, as mentioned in Section 2.2 and shown in Figure 2.1. Figure 2.22 shows the approximate effect of skewing on the cogging torque. Maximum benefit occurs at a skew of one slot pitch, when the end of one slot and the opposite end of the next are aligned along the rotor axis. The amount of skew is in any case limited by the need to maintain the length of the flat-topped section of the flux waveform in the air gap.

Emergency brakes

Servomotors are often fitted with a brake, usually in the front end of the motor between the rotor and the front bearing as shown in Figure 2.23. When the brake operates, a friction disc is pressed against the fixed steel disc visible in the figure. The steel disc is large enough to absorb the heat generated during an emergency stop, without deforming. Two types of brake are used. In one, a spring is restrained by a solenoid during normal operation of the motor and released when operation of the brake mechanism is required. In the other type the duty of the spring is performed by a permanent magnet, the effect of the permanent magnetic field being restrained by the effect of a field from a DC winding. For some applications there must be no backlash in the brake mechanism. A brake with a small amount of backlash may be manufactured at a lower cost than one without backlash, whether operated by a spring or by a permanent magnet. However, the method of manufacture makes the permanent magnet more suitable for the zero backlash type.

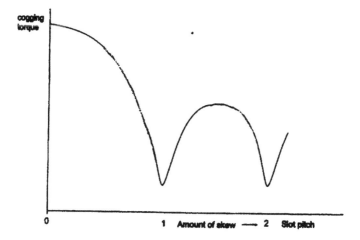

Figure 2.22
Effect of slot skewing on the cogging torque

Both forms of the brake are 'fail-safe', preventing any movement of the motor shaft and connected load following a failure of the power supply to the motor. The brake is also used to hold the motor shaft in position during a shutdown of the work process. In circumstances where the shaft must start exactly from the position reached at the end of the previous operation, the motor must be fitted with the zero backlash form of brake. The brake is not normally used to control the motor during the actual operation. One exception, however, is in a three-dimensional process known as 'two-and-a-half axis' machining, where the product is moved into different positions along a vertical axis by a servomotor, and then machined over two horizontal axes. Here, the motor brake can be used to lock the product into position during the machining process.

Figure 2.23
Emergency brake

CHAPTER 3

BRUSHLESS COMMUTATION

3.1 Introduction

The description of brushless motors as trapezoidal and sinusoidal arises from the shape of the back emfs produced by the stator windings. In this chapter the motors are generally referred to as the squarewave and sinewave types, the main interest being in the form and control of the stator current.

In order to ensure a unidirectional output torque, the current in the stator conductors of the brushless motor must always be in the same direction relative to that of the rotor pole flux. The commutation process has to ensure that the action of switching the direction of the current is synchronized with the movement of the flux in the air gap, and so the motor must have a sensor for measuring the position of the flux wave relative to that of the stator windings. Figure 3.1 shows the current to be supplied by a unit normally known as the motor *drive*, which has the *control circuit* and the *inverter* as its main components. Power is drawn from a DC supply by the drive and supplied to the motor during periods determined from the sensor signals.

The inverter shapes the waveform of the current fed to the motor. Although normally used for a circuit which converts

DC to sinusoidal AC, which is indeed the requirement for the sinewave motor, the name of inverter may also be given to the circuit which supplies the rectangular waveform of current to the squarewave type.

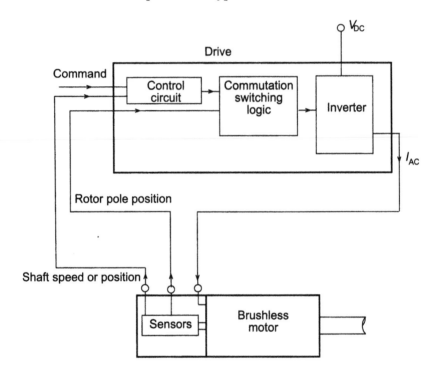

Figure 3.1
Motor drive and current supply

The control circuit manages the speed and position of the motor shaft according to the application. Feedback of information is received from shaft position and speed sensors, allowing the external reference command to be compared with the actual state of the motor. The output from the control circuit is fed through the switching logic circuit to the inverter, where any changes to speed or position are made by adjusting the motor current. The inverter is therefore used to control the magnitude of the stator current, as well as for the formation of its waveform.

3.2 Sensors

This section describes the devices used for the measurement of the rotor pole position, shaft speed and shaft position. The methods of measurement are explained in Section 3.4.

Hall effect (pole position)

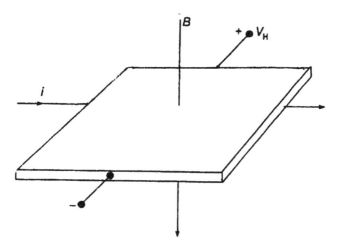

Figure 3.2
The Hall-effect voltage, V_H

The Hall-effect sensor is used to detect the position of the rotor of the squarewave motor. Figure 3.2 shows a current i flowing through a plate of semiconductor material. Voltage V_H is generated by the Hall effect when a magnetic field of density B passes through the plate. If the plate is fixed on the motor stator in a position exposed to the air gap flux, the voltage generated by the Hall effect will be linked to the rotating flux waveform. An advantage of the stator location of the plate is that the generated voltage follows the pattern of the true flux wave in the air gap, including any armature reaction effects. A drawback is that the plate is subject to stator heating and its performance may be affected at high operating temperatures. Overheating is avoided if the sensor is mounted inside the rear housing of the motor, in the path of the flux from an auxiliary magnet which is fixed to the motor shaft. This provides accurate sensing where there is no significant distortion of the flux in the air gap, due to armature reaction.

The brushless AC tacho (shaft speed)

The AC tachogenerator has a three-winding, trapezoidal, brushless structure. It is mounted at the end of the shaft of the squarewave motor in the rear housing, where it generates a set of three-phase voltages which are fed to the drive. The waveforms are identical to the patterns of back emf shown in Figure 2.10, enabling the drive to determine the motor speed from the magnitude of the tacho voltage.

DC tachos

These are the same as AC tachos, except that a circuit board in the rear housing uses the three-phase output to provide a DC signal which is then fed to the drive.

The resolver (pole position, shaft speed and position)

The brushless resolver has a rotor and a stator. The stator carries the input and output windings which are linked by the rotating transformer action of the rotor winding. It is normally fitted to the shaft of the sinewave motor in the rear housing. Figure 3.3 shows a common arrangement where there is one AC input and two output windings which are displaced electrically by 90°. The input winding on the stator forms the primary of a rotating transformer and is usually referred to as the rotor input. The input frequency is typically 5–10 kHz, and AC outputs at the same frequency appear by transformer action at any rotor position. The rotor winding is sinusoidally distributed, with the result that the outputs are amplitude modulated according to the rotor position.

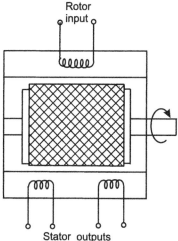

Figure 3.3
Brushless resolver terminals

Digital encoders

The resolver provides the drive with analogue signals, and so digitally controlled drives must incorporate analogue to digital converters. The encoder can provide the drive with signals which are already in digital form. The signals are derived optically from patterned tracks on a disc of translucent material which rotates with the motor shaft. The material can be glass or plastic. Light-emitting and light-sensing devices are placed either side of the disc as shown in Figure 3.4. The light sensors produce the binary 0 or 1 signals when the light is either obscured by the pattern, or transmitted where the disc is exposed. The encoder is driven by the motor shaft in the rear housing, as shown in Figure 3.5.

Incremental encoders

Figure 3.4(a) shows the basic principle of the incremental encoder. As the disc rotates, the light sensor supplies the drive with a signal consisting of a train of pulses. In practice the pattern is also designed to provide the drive with signals sensitive to the direction of rotation of the disc [3]. The drive determines the position of the shaft by adding or subtracting the number of pulses which arrive from either side of a known reference position. With this type of encoder, it is obviously not possible to define the absolute position of the motor shaft without regard to a reference.

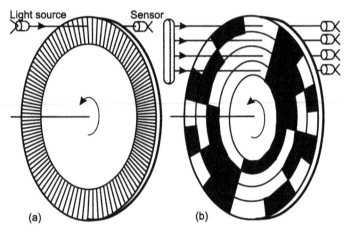

Figure 3.4
Principle of the optical encoder

Figure 3.5
Encoder driven from the rear end of the motor shaft

Absolute encoders

One form of absolute encoder is a development of the optical, incremental type. Unlike the incremental type, however, the signals generated represent discrete shaft positions. Each position is given its own binary number and each digit of the number must have a separate track on the glass disc. Each track has its own scanning, the scans being checked with respect to each other so that reading errors are avoided. The basic principle is shown in Figure 3.4(b). If the disc starts in an anticlockwise direction from the position shown, the shaft positions are defined by the signals $1111 \longrightarrow 0000 \longrightarrow 0001 \ldots \longrightarrow 1110$. The simple four-track encoder shown would be capable of defining only 16 positions of the shaft. A 10-track disc defines 2^{10} or 1024 positions, and so on. Absolute encoders clearly have great complexity and are relatively expensive to produce. The design must combine the photoelectric requirements with a resistance to adverse conditions such as mechanical shock and vibration [4].

There are other designs of absolute encoder. The Sincos type produces sine and cosine signals in analogue form which can be passed directly to the drive, unlike the resolver where a signal from the drive is amplitude-modulated and passed back to the drive. More significantly, the signals from the Sincos device can also be encoded locally and passed to the drive in digital form. Although many times more expensive than a resolver, the Sincos device has clear advantages in the digital control of interconnected servo systems.

3.3 Power electronics

In Figure 3.1, current is supplied to the brushless motor from a power electronic inverter which must repeatedly change the direction of the stator current. The direction cannot be changed without the interruption and reinstatement of a current which may be tens, or even hundreds of amps. There are at least four semiconductor devices which can be used as switches for controlling the direction of currents at such levels:

(a) the thyristor,
(b) the metal oxide, semiconductor field-effect transistor, or MOSFET,
(c) the bipolar junction transistor, or BJT,
(d) the insulated gate, bipolar transistor, or IGBT.

The MOSFET, BJT and IGBT switches close in response to a switching signal input, and open again when the signal is removed. The thyristor switch closes when the switching signal arrives, but remains 'latched' when the signal is removed.

The thyristor

The thyristor was previously known as the silicon-controlled-rectifier, or SCR. Figure 3.6 shows the electrical symbol for the device to have three terminals, known as the gate (G), the anode (A) and the cathode (C). The switch is operated by the application of a gate current I_G which has a low value compared to the switched power current I_A. Current I_G is normally applied in short pulses. Figure 3.6(a) shows V as a constant voltage. Current I_A switches on as soon as the first gate pulse arrives, and then stays on as the switch latches. In Figure 3.6(b), V is alternating and so the current is turned off naturally when V falls to zero, or soon after depending on circuit inductance.

The best uses for the thyristor lie in applications where operational advantage can be taken of natural current zeros, such as rectification of an AC supply. Figure 3.6(b) is an example of such rectification where the average value of the power current is positive and of a magnitude which can be controlled by the position of the gate pulses relative to the supply voltage zeros. We have seen that the needs of the brushless motor are quite different. The stator current is derived from a DC source and must be repeatedly switched on and off at moments dictated by the position of the rotor. Extra circuitry is needed to force the thyristor current down to zero, with the result that overall switching times are too long for accurate commutation of the brushless motor.

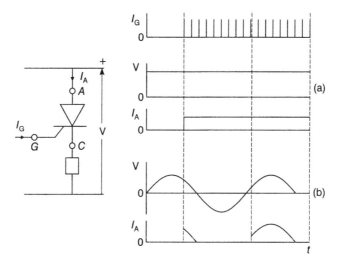

Figure 3.6
Thyristor control of direct and alternating currents

Gate turn-off thyristor

The gate turn-off (GTO) thryristor has a modified structure which allows the device to be turned off by extracting a pulse of current from the gate. The GTO must be provided with an initial pulse of gate current to effect turn-on, after which a small gate current must be maintained throughout the required period of power current conduction. Although the GTO was a relatively early development, advances in other directions in semiconductor technology have obviated its use in brushless motor drives.

The power MOSFET

The circuit symbol of the MOSFET is shown in Figure 3.7. The switched current flows between the drain (D) and source (S) terminals, falling to zero when the gate signal is removed. The power version of the MOSFET is used where advantage can be taken of its very low gate current and high frequency switching features, for example in switched mode power supplies. The main limitation is that at a fixed manufacturing cost, the resistance of the MOSFET during the 'switched-on' periods rises according to its designed operating voltage. This means that it is normally used in applications which have operating voltages lower than the levels used for industrial brushless servomotors.

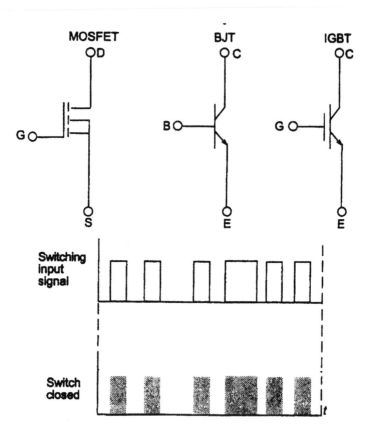

Figure 3.7
Semiconductor power switches

The power BJT

Technical development of the brushless motor did not really move quickly until the 1970s, with the commercial development of the bipolar junction transistor, or BJT. Since then, development of the high power BJT has gone hand in hand with the dramatic improvements in the design and performance of the brushless servomotor. Modern semiconductor switches based on the BJT are well able to cope with the needs of industrial motion control.

Figure 3.7 shows the circuit symbol for the transistor to have three terminals, the base (B), the emitter (E) and the collector (C). In microelectronic, analogue applications the transistor is often used as an amplifier. On the other hand, the power transistor is used almost entirely as a switch. Switched current flows between the collector and emitter terminals in response to current signals fed to the base. Compared to a MOSFET of equivalent cost, the power BJT has a low resistance to the flow of high currents at high operating voltages. A drawback is that the base current must commence and remain at a level of several amps if (using the mechanical analogy) the switch is to be kept firmly closed and without significant 'contact' resistance.

The IGBT

The very low power requirements of the gate of the MOSFET, and the BJT characteristic of low resistance at high current and high voltage, are brought together in the IGBT, which is sometimes referred to as a MOS-gated BJT. The circuit symbol is shown Figure 3.7. The device is used widely in brushless motor drives, although it does suffer from a high turn-off time in comparison with the MOSFET and must be applied carefully where high switching frequencies are needed. The design of the device varies according to the broad band of operating frequency required, the central band being from 1 to 10 kHz.

To summarize, it can be said that the IGBT is much easier than the power BJT to drive, but has a lower switching speed compared to the MOSFET. The power MOSFET has a high switching speed, but has a relatively low switched current capability at the higher voltages. The BJT normally has the lowest cost, followed in order by the IGBT and the power MOSFET.

Power electronic inverters

A single semiconductor switch from the types described above can be used to turn a current on or off for current flow in only one direction. Two switches are needed for a current which must be turned on and off in both directions.

Single phase

The single-phase inverter is not normally used to supply the industrial servomotor, but it does provide a simple introduction to the principle of operation of the three-phase circuit. Figure 3.8 shows a single-phase brushless motor connected to a half-bridge single-phase inverter. The rectangular waveform of current which must be supplied to the motor is shown at the top of the diagram. The input signals to the switches are derived from the pole-position sensor and supplied through the switching logic circuit, as shown in Figure 3.1. The three-wire DC supply provides voltages of $\pm V/2$.

For (a), θ is between 0 and 180°. Q_1 is on, Q_2 is off and a current flows in the motor stator in the direction shown. In (c), θ is between 0 and 180°. Q_2 is on and Q_1 is off and so current flows in the opposite direction through the stator to that in (a). The two switches must never be allowed to be on together, otherwise the supply would be short-circuited, and so Q_1 must turn off before Q_2 turns on and vice versa. However, the stator winding has inductance and so the stator

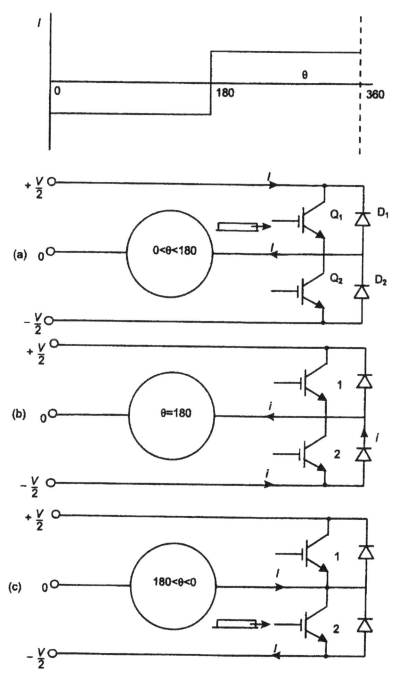

Figure 3.8
Half-bridge, single-phase inverter

current must be allowed to flow along an alternative path while both switches are off. After Q_1 turns off, the stator current circulates through 'freewheeling' diode D_2 around the path shown in (b) and dies away. The circulating current flows towards the positive supply voltage and so the power source must be capable of accepting the associated flow of energy. Similar currents and energies flow around the upper half of the circuit after Q_2 turns off and before Q_1 turns on.

The half-bridge circuit makes poor use of the power source as only half the supply voltage appears across the load. The full-bridge circuit of Figure 3.9 has no such drawback, although it does need two extra semiconductor switches. Current flows through the motor along the path shown when Q_3 and Q_4 are on, and reverses when the conducting pair changes to Q_1 and Q_2. All switches must be off during the changeover, leaving the current to die away through D_1, D_2 and the power supply, which must be able to accept reverse current.

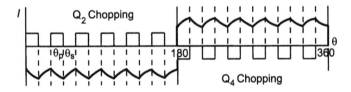

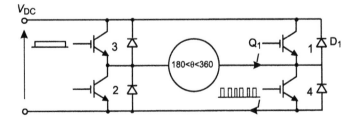

Figure 3.9
Full-bridge, single-phase inverter with PWM regulation

Full-bridge three-phase circuit

The inverter circuit for the three-phase brushless motor is similar to the full-bridge single-phase circuit except that it has three pairs of semiconductor switches and three pairs of diodes. The details are covered in Section 3.4.

Control of current magnitude

Power electronic inverters convert direct current into alternating current of rectangular or sinusoidal form. They are also used to control the magnitude of the current and hence the motor torque. In the usual method, the current is 'chopped' by one or both of the pair of conducting switches. For example, in Figure 3.9 the lower switch of the Q_3–Q_4 pair is repeatedly turned on and off during the conduction period. The switching signal comes through the switching logic circuit from the control circuit in Figure 3.1. When Q_4 turns off, the motor becomes disconnected from the voltage supply. The motor current then decays around the D_1–Q_3 path at a rate determined by the resistance and inductance of the stator winding. When Q_4 turns on again, the supply voltage is reconnected and the current rises.

PWM

In Figure 3.9, the average current is controlled by PWM or *pulse width modulation*. As the pulse width θ_P is increased, the average current rises. The current is maximum when the space $\theta_S = 0$, and zero when $\theta_P = 0$. To avoid undue ripple in the current waveform, $\theta_P + \theta_S$ is normally 10% or less of the half-cycle angle, or 18 electrical degrees. For example, the rotor of a four-pole squarewave motor running at 6000 rpm, or 1 revolution (720 elec°) every 10 milliseconds, would take $10 \times 18/720 = 0.25$ ms to pass through 18 electrical degrees. The PWM switching frequency required would be at least 4 kHz.

Current regulation

In the PWM method the current is controlled by adjusting the width of current pulses which occur at a fixed rate. In the *current regulation* method, the chopping switch is turned on or off if the current is too low or too high respectively. The magnitude is measured continuously by means of a current transducer and switching occurs at the limits of a narrow band around the average value required (Figure 3.10).

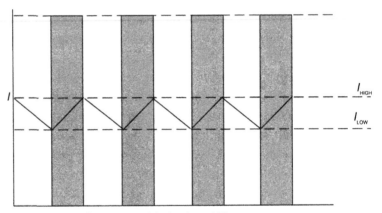

Current-regulated pulse widths

Figure 3.10
Current regulation

In the simplest case, switching occurs at an irregular rate, depending on fluctuations in the current requirement. In an alternative current regulation method, the irregularities are avoided by preventing the switch from turning on again until the end of a switching period of a fixed length.

3.4   Three-phase commutation

Introduction

The block diagram of Figure 3.1 shows the main components of the brushless motor drive. The diagram also shows the speed and position sensors in the rear housing of the motor. In Section 3.2 we looked at the various types of sensor used for the generation of signals which are dependent on the speed and position of the motor shaft, and the position of the rotor poles. The drive shown in Figure 3.1 is assumed to develop digital commutation signals. Any analogue signals from the sensors must be passed through an A/D converter, either within the drive or locally at the motor. The subject of Power Electronics was introduced in Section 3.3, and the operation of the power electronic inverter was described by taking the single-phase case. The work covered the simple commutation of the motor current using signals from a pole position sensor, and also the basic methods of control of the current magnitude.

In this section we will see how the commutation of the motor current and the speed and position of the shaft are controlled on a three-phase basis. There are two areas of interest:
1. the operation of the three-phase inverter,
2. the methods of three-phase measurement of pole positions and shaft speed and position.

Operation of the full-bridge three-phase inverter

The general layout of the inverter often used for the commutation of the squarewave or sinewave motor is shown in Figure 3.11. It is a simple extension of the full-bridge single-phase circuit of Figure 3.9. Only two phases of the squarewave motor require a current supply at any one time. The upper half of Figure 3.12 indicates the switches which must conduct at each stage of the flow of phase currents to the squarewave motor from the inverter of Figure 3.11. For example, when $\theta = 30°$, Q_1 and Q_2 are on and current flow is A⟶C and when $\theta = 270°$, Q_5 and Q_6 are on and current flow is C⟶B.

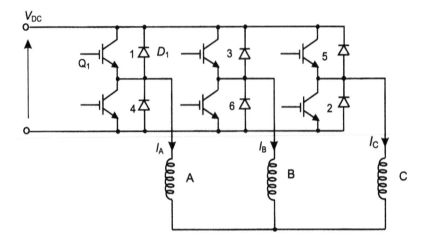

Figure 3.11
Inverter for the three-phase squarewave and sinewave motors

The sinewave motor requires a continuous current supply to all three phases, the current in each being zero only at the zero-crossing points of the sinusoid. Three switches must conduct at any one time. The lower half of Figure 3.12 shows the required sequence of conduction of the inverter switches for the sinewave motor, as the angle of the rotor changes.

Current magnitude

There are several ways in which voltage PWM or current regulation may be applied to the three-phase full-bridge inverter. The current in each phase of the squarewave motor flows for 120 electrical degrees and so the upper switches in Figure 3.11 can conduct continuously and the lower ones can be interrupted over the full 120°, as in the single-phase circuit of Figure 3.9. On the other hand, the current in each pair of phases may be seen in Figure 3.12 to flow for only 60°, and alternative switching strategies are possible where 120° chopping of each phase results from operation of the chopping switches over only 60°.

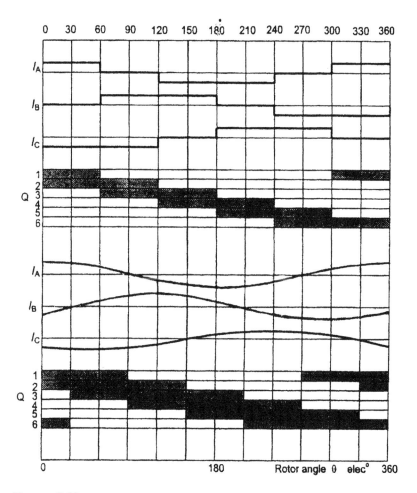

Figure 3.12
Inverter switching of squarewave and sinewave currents

Commutation of the Squarewave Motor

Measurement of pole position

The Hall-effect sensor is used to detect the rotor pole positions at which the direction of each phase current must be reversed. The current in each phase of a squarewave motor must be individually controlled and so three sensors are needed. These may be mounted evenly at the 120 electrical degree intervals of a set of three-phase windings. However, at least one manufacturer mounts the sensors over an angle of only 120 of the 360 electrical degrees, giving a spacing of 60°. For a four-pole motor, the 120 electrical degree zone for the sensors would be 60 mechanical degrees. Figure 3.13 shows the outputs p, q, r from position sensors mounted on a two-pole motor over

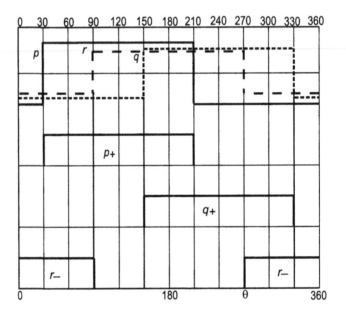

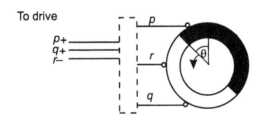

Figure 3.13
Outputs from Hall-effect sensors

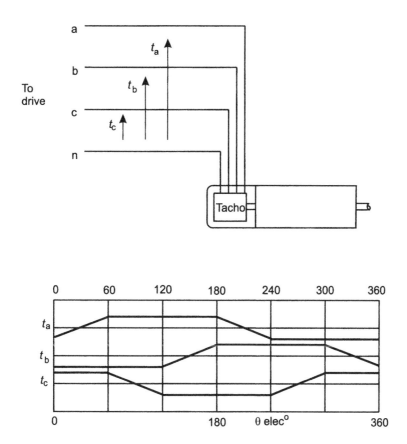

Figure 3.14
AC tacho outputs to the drive

an angle of 120°. The rotor angle θ is given in mechanical degrees, which are the same as electrical degrees for the two-pole motor. Selection of the positive half-cycles from p and q, and the inverted, negative half-cycle from r, gives the three correctly spaced signals shown in the diagram. These are fed to the drive where they are used to trigger the commutation.

Measurement of shaft speed

The speed of the motor is related to the magnitude of the alternating back emf generated by each pair of phase windings of the Y-connected motor by the expression

$$E = K_E \omega$$

where K_E is expressed as line-to-line volts per rad/sec. The emf of the loaded motor is of course inaccessible as a means of measuring the speed. Instead, the output from an AC tacho is used by the drive to produce a signal proportional to the line-to-line emf of the motor, and therefore to its speed. Figure 3.14 shows the outputs from a two-pole Y-connected brushless tacho which is fitted to a two-pole Y-connected motor. For a four-pole motor fitted with a four-pole tacho, the complete electrical cycle of 360° would occur over an angle of 180 mechanical degrees.

Figure 3.15(a) shows the three rotor position signals. The tacho output is typically 10 V/krpm and is reduced in the drive to a manageable level. When the signals are used by the drive to commutate the tacho phase emfs, which are rectified at the same time, the result is the DC voltage shown in 3.15(b). The voltage is made up from the top sections of each half-cycle of each emf waveform, drawn with rounded rather than ideal flat tops to give the diagram clarity. For example,

when $\theta = 30°$, t_c turns off and rectified t_b turns on, using p_+

and when $\theta = 90°$, rectified t_b turns off and t_a turns on, using r_-.

The DC voltage in Figure 3.15(b) is proportional to the speed of the brushless motor at all rotor angles, as can be seen by comparison with the line-to-line emfs (Figure 3.15(c)). The ideal flat tops are again rounded in the diagram and only one emf is shown as a complete waveform:

$$E_{ca} = E_c - E_a$$

The waveform of the rectified, commutated, per-phase emfs from the tacho is seen to be synchronized with the envelope of the (rectified) line-to-line emfs of the motor.

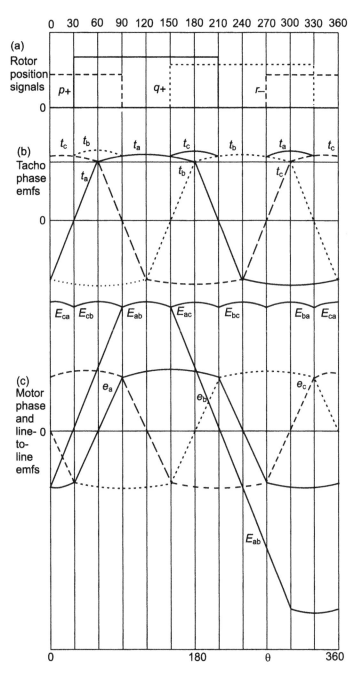

Figure 3.15
Measurement of shaft speed of the trapezoidal motor

Measurement of shaft position

Each phase current of the squarewave motor is commutated at the beginning and end of 180 electrical degrees of the pole rotation. No information about intermediate positions is given to the drive by the Hall plate or tacho sensors and so nothing is known about the position of the motor shaft between the commutation points. If a squarewave motor is to be used in an application which requires position and speed control, it can be fitted with an incremental encoder in addition to the tacho.

Commutation of the sinewave motor

In the ideal squarewave motor, the phase current is independent of the rotor angle between the commutation points. PWM is applied as a means of altering the magnitude of the rectangular current waveform. In the sinewave motor, the current must depend on the sine of the rotor angle and so PWM is used to produce the sinusoidal shape of the waveform as well as to control its magnitude.

The two conducting supply lines of the squarewave motor carry the same current. For the sinewave motor, however, Figure 3.12 shows the inverter output currents in general to have different magnitudes at all rotor angles. This means that PWM must be applied separately to each line current, and also that the rotor angle of the sinewave motor must be known at all times.

Measurement of rotor angle

The sinewave motor is often fitted with a brushless resolver which continuously monitors the position of the rotor. As the position is known at all times, separate shaft speed and position sensors are not needed. Figure 3.16 shows the two outputs which are fed to the drive from the resolver. They consist of the resolver input signal, modulated according to the sine and cosine of the rotor angle in mechanical degrees.

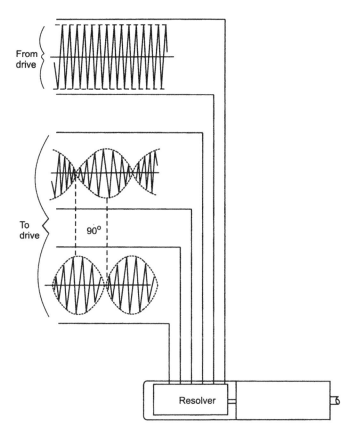

Figure 3.16
Resolver signals

A positive increase in angle normally indicates clockwise rotation of the motor shaft, looking at the drive end. The actual direction of rotation is found from the phase sequence of the two signals. Figure 3.17 shows how the two signals are compared. The drive is able to determine the quadrant of the angular position as only one is common to the sine and cosine measurement.

The resolver outputs occur at the same frequency as the phase emfs only in the case of a two-pole motor, where the mechanical and electrical rotor angles coincide. Figure 3.18 shows the ideal emfs generated by the a-phase of four-, six-,

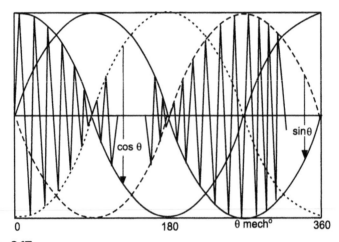

Figure 3.17
Definition of shaft position

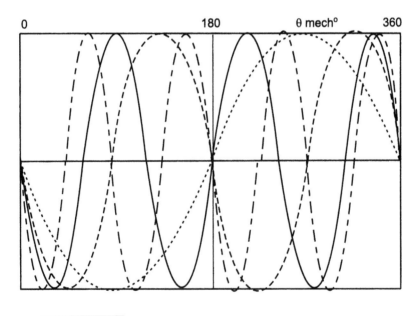

Figure 3.18
Mechanical positions of rotor poles

and eight-pole motors as the rotors turn through 360 mechanical degrees. The figure also shows the signal which originates from the resolver for the sine of the mechanical angle of the rotor. The drive converts the mechanical angle determined from the resolver signals into the electrical angular position required for commutation, according to the number of rotor poles.

3.5   The motor-sensor combination

There are two types of brushless motor and several types of sensor, giving many possible combinations. The combination used in practice is chosen partly according to its cost-effectiveness for the application in hand. The choice is also affected by the potential scale and future development of the application. Sinusoidal drive systems are generally more expensive than the trapezoidal type, and there is a great difference between the price of an absolute and an incremental encoder. In general, the least expensive sensors are used with trapezoidal motors.

Trapezoidal motor–Hall-effect sensor

The speed of the trapezoidal motor can be estimated by using the signals from the Hall-effect sensor already fitted for the purpose of commutation. The signals are too widely spaced to be used for position control, but speed regulation above about 500 rpm is possible. The method is limited mainly to on–off, fixed speed applications, one example being the brushless industrial fan motor. In this example, the rival pricewise is the induction motor and inverter unit.

Trapezoidal motor–tacho and incremental encoder

The shaft speed sensor of the trapezoidal motor is normally in the form of a tacho, and an incremental encoder is added when measurement of the shaft position is required. The encoder output can be processed to give a measurement of motor speed, but the resulting signal is not usually good enough at the mid to low end of the range. This is due to the presence of a ripple which is inversely proportional to speed. The ripple may be reduced by increasing the number of lines on the encoder, but this increases the cost and also limits the maximum encoder speed. Low speed control of the trapezoidal motor is in any case subject to the effect of its cogging torque. The encoder signal can, however, give accurate positioning of a load, particularly where the load movement is reduced in comparison to that of the motor shaft through a transmission mechanism. A pulley and belt driven packaging machine, for example, may easily be controlled to within 0.5 mm.

Sinusoidal motor—resolver or absolute encoder

This combination is used for applications which demand very high accuracy of load speed and position. Many such cases occur in the control of machine tools such as lathes and milling machines. Choosing between the resolver and the absolute encoder on the basis of cost tends to be unrealistic. The drive accepts the signals from an absolute encoder in digital form, but must first process the analogue signals from a resolver. Even if a resolver is only 10% of the price of a Sincos absolute encoder, the overall costs of drive plus sensor may not differ significantly. The absolute encoder can be the best choice for relatively large scale installations where an integrated approach is taken to the digital control system.

CHAPTER 4

MOTOR AND LOAD DYNAMICS

4.1 Introduction

The aim of this chapter is to study the electrical and dynamic characteristics of the loaded motor, and the effects of the interplay between the motor and load masses. The emphasis is on understanding the characteristics of the motor and load as one component of the overall system, rather than on the detailed methods of system control which are well explained in existing text [2], [4]. The treatment of servomotor control theory is limited here to a very short introduction to the main principles.

The electrical time constant of the brushed motor is often ignored, but this is generally not possible in the brushless machine due to its relatively high inductance. The main part of this chapter starts by developing the basic equations for the brushless motor in terms of the mechanical and electrical time constants. An introduction is given to the method of Laplace transformation, which is then used to study the effects of the time constants on the electromechanical behaviour of the motor.

When the brushless motor is used for incremental motion, the motor rating may be affected by the relative values of the motor and load inertias. Here, we look to the optimization of the design of the transmission mechanism between the two masses. A gear train, belt and pulley or ball screw drive can each be designed in a way which keeps the drive motor heating to a minimum. Such transmission mechanisms can be applied in ways which differ widely, and it is unrealistic to describe any numerical example as typical. Some examples are included here in order to illustrate the main principles involved in the efficient connection of the motor to the load.

The basic analysis of the performance of a loaded motor usually assumes that the shafts and couplings between the rotor of the motor and the load are completely rigid. In practice, flexibility of the motor–load connection may alter performance and cause problems with the control system. Flexibility in the connection is normally known as *compliance*. The electromechanical effects of the compliance of the connection between the motor and the load are treated at the end of the chapter.

Note that throughout this chapter, the speed of the motor is denoted by ω_m instead of the symbol ω used previously.

4.2 Motor control

In Chapter 3 we studied the inverter, which is one of the two main components of the drive. The drive also has the task of controlling the motor shaft position and speed, using input command signals and signals from the respective sensors. The motor, sensor and drive form a closed circuit normally known as a closed loop.

Closed loops

At a fixed supply voltage, the voltage drop across the winding resistance of a permanent magnet brushed motor causes the speed to fall as the load torque rises. In the brushless motor, the speed is affected by voltage drops which occur across the winding inductance as well as across the winding resistance. In the ideal case, the requirement of most motion control applications is for the motor speed to be independent of everything except the speed command. There is therefore the need for *feedback* of information about the motor speed from moment to moment, so that the motor input current may be correctly and continuously adjusted.

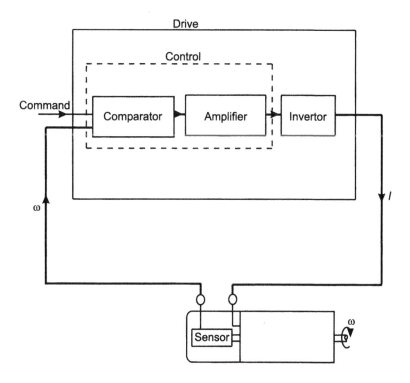

Figure 4.1
Closed-loop speed control

The basic blocks of the control unit are shown as a comparator and an amplifier in Figure 4.1. The closed loop is made up of the sensor, control unit, inverter output circuit and of course the motor. The difference between the sensor output and the desired reference input is transmitted as an error signal from the comparator to the amplifier. The drive output is then adjusted at the inverter according to the amplifier output so as to reduce the error in speed.

Speed regulation

Figure 4.2 shows the four possible combinations of positive and negative velocity and torque as a four-quadrant diagram. The shaded area of Figure 4.2(a) indicates that the control system is designed for only positive velocity and torque, and the unidirectional system is normally known as a *speed regulator*.

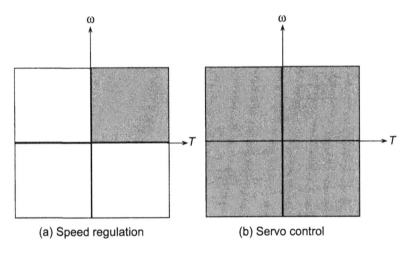

(a) Speed regulation   (b) Servo control

Figure 4.2
Motor torque and velocity

Servo control

A servo system controls the direction of rotation as well as the torque direction of the motor, and so the motor is operated over all four quadrants of the velocity–torque plane shown in Figure 4.2(b). For many applications there may also be the need for control of the position of the load, and therefore of the motor shaft. Figure 4.3 shows the additional feedback loop required for position control.

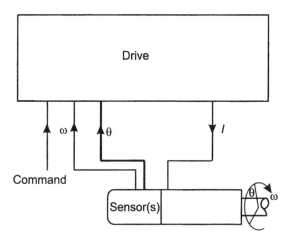

Figure 4.3
Closed-loop control of velocity and position

Of major importance in the design of the control system is a knowledge of the 'open-loop' properties of the motor and load unit, before its inclusion in the closed loop. Sections 4.3 and 4.4 look at the characteristics of the motor and load, and at how they affect the open-loop behaviour.

4.3　Motor equations

Equations were developed in Chapter 1 for the steady-state characteristics of the brushed motor. It was seen in Chapter 2 that one difference between the brushed and brushless motors is that one motor incorporates a mechanical inverter consisting of brushes and commutator whereas the other has an external, power electronic inverter. Both machines operate from a DC supply.

Constant speeds

For the brushed motor, the current is subjected to commutation in only a small part of the winding at any one time. The distorting effect of the inductance of the part under commutation is low in comparison to the smoothing effect of the inductance of the remaining winding. The direct current to the brushed motor is therefore largely unaffected by the commutation process, and the speed at steady-state is usually assumed to be independent of motor inductance. Such an assumption cannot be made for the brushless motor, where commutation occurs at the same moment for a complete winding. Any formulation of steady-state equations must take account of the voltage drop across the winding inductance as well as that across the winding resistance. Transient demands of torque and speed are, however, the common requirements for a brushless servomotor, and steady-state equations are unlikely to be of use.

Speed variations

An equation for the motor speed under transient conditions must take account of all mechanical and electrical factors which affect a change in speed. The rate of change of the motor torque is limited by the rate at which the motor current can be changed, which is in turn limited by the motor inductance. The inductance can be found by applying a sinusoidal AC voltage of angular frequency ω rad/s to the stator, after first locking the rotor shaft in a fixed position. The line-to-line impedance of the motor to the flow of alternating current is shown in Figure 4.4 to consist of resistance R and the motor reactance ωL, where L is the electrical inductance of the motor. Resistance R_M accounts for the power losses in the magnetic circuit but as its value is normally high in comparison with ωL, its effect on the overall circuit is usually ignored. The voltage applied across the lines is

$$V_{\text{rms}} = I_{\text{rms}}(R + j\omega L)$$

L is the only unknown and is normally assumed to have the same value over a wide range of frequency. Figure 4.5 shows an equivalent of the stator input circuit, which consists of the line-to-line back emf, inductance and resistance. We will study how quickly the motor speed can be changed on the assumption that the input voltage V is applied suddenly, as a step input.

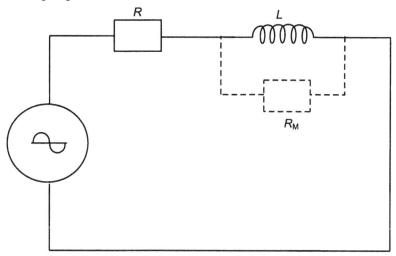

Figure 4.4
Locked rotor equivalent circuit

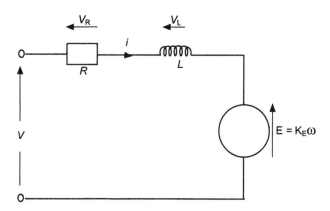

Figure 4.5
An equivalent circuit for the brushless motor

The electrical equation

In Figure 4.5, the volt drop across L acts in the direction shown when the rate of change of current is positive. In other words, V_L opposes the change in current. The *electrical equation* of the motor is seen to be

$$V = L\frac{\mathrm{d}}{\mathrm{d}t}i + Ri + K_E\,\omega_m$$

where V is the applied voltage and i is the current at time t.

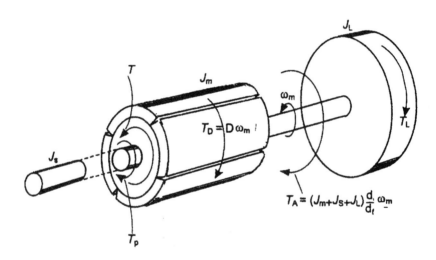

Figure 4.6
The dynamic system

The dynamic equation

The rate at which the motor speed can change is clearly affected by the moment of mechanical inertia of the driven load, and also by the moments of inertia of the rotor and sensor. The unit of the moment of inertia is the kgm². In Figure 4.6, the rotor of a motor of inertia J_m is connected to a load mass of inertia J_L and to a sensor of inertia J_S. The torque T

produced by the motor is opposed by the torque T_f, due to bearing and bearing seal friction, and by the torque T_D due to the viscous damping from iron losses and windage. The motor must also react to the torque T_A imposed by the mass of the motor, load and sensor during speed changes. The *dynamic equation* is

$$T = J\frac{\mathrm{d}}{\mathrm{d}t}\omega_m + D\omega_m + T_f + T_L$$

where $J = J_m + J_L + J_S$. At constant speed, the expression reduces to the steady-state form.

Electrical and mechanical time constants

Following the application to a stationary motor of a constant voltage in the form of a step input, the rotor speed and stator current each rise over time. Suppose that the rotor is locked in position and stationary throughout. With ω_m set to zero, the electrical equation becomes

$$V = L\frac{\mathrm{d}}{\mathrm{d}t}i + Ri$$

Dividing through by R gives the final stator current as

$$I = \frac{L}{R}\frac{\mathrm{d}}{\mathrm{d}t}i + i$$

Solution of this expression shows that the current rises exponentially towards its final value according to

$$i = I(1 - e^{-Rt/L})$$

Figure 4.7(a) shows the form of the current rise following the application of the step input of voltage. When $t = L/R$, the current reaches $100(1 - e^{-1})\%$, or 63.2% of its final value. The *electrical time constant* is defined as

$$\tau_e = \frac{L}{R}$$

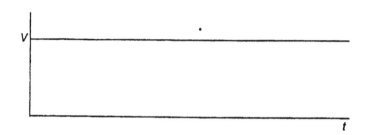

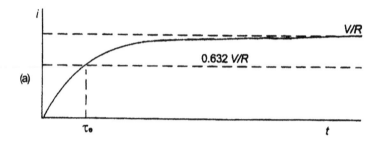

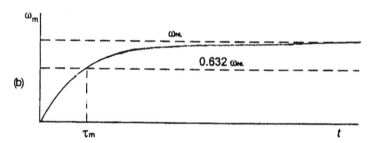

Figure 4.7
(a) Rise of current with rotor locked (b) Rise of speed when $L = 0$

Suppose now that the rotor is stationary but free to rotate, with no load and no supply voltage. Following the application of the step-input voltage to the stator, current flows into the stator winding and the rotor accelerates. If the opposing torques due to viscous damping and friction are assumed to be insignificant, the accelerating torque on the rotor is found from the dynamic equation to be

$$T = J_m \frac{d}{dt} \omega_m$$

where J_m is from this point taken to include the inertia of the sensor. The rate of rise of stator current, and therefore of torque and rotor speed, is affected by the electrical time constant. In order to study the other factors which affect the rate of speed rise after the sudden application of V, assume for the moment that the motor has no inductance and therefore an electrical time constant of zero. The electrical equation reduces to

$$V = Ri + K_E\omega_m$$

The final no-load speed at voltage V would be

$$\omega_{NL} = \frac{V}{K_E}$$

Replacing T by $K_T i$ and combining the last three equations above gives

$$\omega_{NL} = \frac{RJ_m}{K_T K_E}\frac{d}{dt}\omega_m + \omega_m$$

Solving the last expression shows the speed of the unloaded motor with no inductance to rise with time according to

$$\omega_m = \omega_{NL}(1 - e^{-t/\tau_m}) \quad \text{where} \quad \tau_m = \frac{RJ_m}{K_T K_E}$$

In Figure 4.7(b), the speed reaches 63.2% of its final value when t equals the *mechanical time constant* τ_m. Resistance R places the limit on the current and torque for a motor with no inductance, which accounts for the appearance of the electrical resistance in the mechanical time constant. A hypothetical motor with $R = 0$ and $L = 0$ would reach full speed at the instant of application of the supply voltage in response to an infinite current impulse. The rate of rise of speed of a real motor is, of course, subject to the combined effects of the electrical and mechanical time constants.

The electrical time constant of a brushed motor is usually low compared to its mechanical constant, and analysis is often eased by ignoring the motor inductance. This simplification cannot be used for the brushless motor, where in many cases $\tau_m < \tau_e$. Taking, for example, the trapezoidal motor in Table 4.1,

Table 4.1   Specification of a four-pole brushless servomotor

Motor type		Trapezoidal	Sinusoidal
Line-to-line resistance	$R\,\Omega$	3.5	3.5
Torque constant	K_T Nm	0.84	1.02
Max. continuous current	I_S A	5.4 DC	4.4 rms
Max. current (peak)	I_M A	26	26
Max. speed	ω_M rpm	6000	6000
Max. volts (peak)	V_M V	530	530
Inductance	L mH	24	24
Rotormoment of inertia	J kgm^2	0.00028	0.00022
Mechanical time constant	τ_m ms	1.4	1.4
Thermal time constant	τ_{th} mins	35	35
Thermal resistance	°C/W	0.77	0.77

$$\tau_e = \frac{L}{R} = \frac{24.0}{3.5} = 6.9 \text{ ms}$$

and

$$\tau_m = \frac{RJ_m}{K_T K_E} = \frac{3.5 \times 0.00028}{0.84 \times 0.84} = 1.4 \text{ ms}$$

The electrical time constant of this motor is therefore about five times the mechanical value. When the supply voltage is switched on, the rate of acceleration of the rotor mass would clearly be affected by the opposition of the voltage $L di/dt$ to the build-up of motor current, as well as by the moment of inertia of the rotor

The electromechanical equation of the unloaded motor

Assume now that we have a stationary motor which is free to rotate, and which has inductance, resistance and inertia. A step input of voltage V is applied at $t = 0$. We already know that the electrical and mechanical equations for the unloaded motor are

$$V = L\frac{\mathrm{d}}{\mathrm{d}t}i + Ri + K_E\omega_m$$

and

$$T = J_m \frac{\mathrm{d}}{\mathrm{d}t}\omega_m$$

Viscous damping and friction are assumed to be negligible. Replacing T by $K_T i$ and combining the last two equations above gives

$$V = \frac{LJ_m}{K_T}\left(\frac{\mathrm{d}^2}{\mathrm{d}t^2}\omega_m + \frac{R}{L}\frac{\mathrm{d}}{\mathrm{d}t}\omega_m + \frac{K_T K_E}{LJ_m}\omega_m\right)$$

Using the electrical and mechanical time constants already developed in terms of R, L, J_m, K_T and K_E, the last equation above may be written as

$$V = K_E\left(\tau_e\tau_m\frac{\mathrm{d}^2}{\mathrm{d}t^2}\omega_m + \tau_m\frac{\mathrm{d}}{\mathrm{d}t}\omega_m + \omega_m\right)$$

This equation can be used to study the response over a commutation period. For the brushed motor, the equation reduces at steady state to the familier form $V = K_E\omega_m$

Frequency response

Up to this point, we have dealt with the motor speed response to a step input of voltage as a transient disturbance in terms of the function of time $\omega_m(t)$. We have seen that the speed would rise exponentially for a hypothetical motor with no inductance. However, it is possible for a sinusoidal element to be introduced into the transient response of the motor speed, in the practical case when both inductance and inertia are present. The actual shape of the waveform of motor speed against time then depends on the frequency of the sinusoid, as well as on the exponential change in the magnitude of the response. Both features can be displayed on a four-quadrant diagram known as the s-plane.

The s-plane

The s-plane is shown in Figure 4.8. For linear systems the upper and lower quadrants form mirror images, and the points of such a plane are defined by

$$s = \sigma \pm j\omega$$

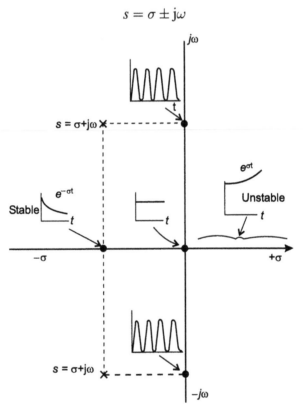

Figure 4.8
The s-plane of $f(t)$

A purely sinusoidal part of the speed response has a frequency of ω and appears on the $j\omega$ axis, and also at the conjugate position $-j\omega$. The point has a co-ordinate along the negative horizontal axis which is determined by the rate of the exponential decay of the response. Any exponentially rising and therefore unstable component appears on the positive horizontal axis. The steady-state component appears at the origin, where $s = 0$. In Figure 4.9, the response with a noticeable oscillatory component is positioned relatively close to the $j\omega$ axis. Only a small overshoot of speed occurs at the 45° position, and no overshoot occurs at the lowest position shown where the frequency of the sinusoid is relatively low and the exponential decay of its magnitude is relatively large. The question which now arises is how do we find the position on the s-plane of a response which is given as a function of time.

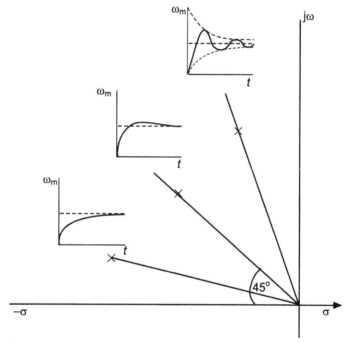

Figure 4.9
Speed responses associated with various points on the s-plane

The Laplace transformation

The Laplace approach allows us to represent time-varying functions on the s-plane. The Laplace transform of $f(t)$ is given by

$$f(s) = \int_{0}^{\infty} \frac{f(t)}{e^{st}} \, dt$$

Note that $f(s)$ is not defined for $t \le 0$. In the integral, e^{st} is dimensionless and so s has the dimension of t^{-1}, that is of frequency. The Laplace integration is normally done from a list of standard transforms. Five examples are shown in Table 4.2. The first four are known as functional transforms and are found from the Laplace integral given above, using the normal method of integration. The first line of the table shows the transform of a step function. The step function has a value of zero for $t \le 0$ and has a constant value of V for $t > 0$.

Table 4.2   Laplace transforms

Function of time	Laplace transform
V (step)	$\dfrac{V}{s}$
t (ramp)	$\dfrac{1}{s^2}$
$e^{-\sigma t}$	$\dfrac{-1}{s+\sigma}$
$\cos \omega t$	$\dfrac{s}{s^2 + \omega^2}$
$\dfrac{d}{dt} f(t)$	$sf(s)$

The last line of the table shows an operational transform, developed using integration by parts. It states that the transform of the rate of change of a time-varying function is simply equal to s multiplied by the transform of the function. For example, the Laplace transform of the rate of change with time of motor speed is

$$f(s) = s\omega_m(s)$$

where $\omega_m(s)$ is the Laplace transform of the motor speed $\omega_m(t)$. It follows that the transform of the second order of the rate of speed change is $s^2\omega_m(s)$, and so on.

The poles of the frequency response

Let us now return to the case of the unloaded motor which has inductance and inertia, and which is supplied with step voltage V at $t = 0$. The electromechanical equation has been shown to be

$$V = K_E \left(\tau_e \tau_m \frac{d^2}{dt^2} \omega_m + \tau_m \frac{d}{dt} \omega_m + \omega_m \right)$$

Applying Laplace transforms gives

$$\frac{V}{sK_E} = \tau_e \tau_m s^2 \omega_m(s) + \tau_m s \omega_m(s) + \omega_m(s)$$

Rearranging, we obtain

$$\omega_m(s) = \frac{V}{K_E s(s^2 \tau_e \tau_m + s \tau_m + 1)}$$

The last expression above gives the speed response of the motor to a step input. Values of s which make the denominator equal to zero are known as the system *poles*, defining the composition of the response on the s-plane. The pole at $s = 0$ represents a steady-state part of the transform.

Transfer functions

A transfer function is written for a system output relative to a system input. There is a separate transfer function for each combination of input and output. In Figure 4.10 the system input is the voltage $V(s)$ and the output is the motor speed $\omega(s)$. The figure shows the transfer function of the unloaded motor as

$$\frac{\omega_m(s)}{V(s)} = \frac{1}{K_E(s^2 \tau_e \tau_m + s \tau_m + 1)}$$

The denominator of the transfer function may be used to find the poles of the transient speed response to the step voltage input. The steady-state pole at $s = 0$ does not appear in the transfer function. The trapezoidal motor specified in

Table 4.1 has already been shown to have a ratio of electrical to mechanical time constant of approximately 5:1, and so we would expect its speed response to a step input to be dominated by the effect of inductance. The numerical values of the motor constants are

$$K_E(= K_T) = 0.84, \quad \tau_e = 0.0069, \quad \tau_m = 0.0014$$

Using these values in the above transfer function gives

$$\frac{\omega_m(s)}{V(s)} = \frac{1}{8.1 \times 10^{-6}s^2 + 11.8 \times 10^{-4}s + 0.84}$$

The poles are defined when the denominator of the transfer function equals zero. Solving the quadratic in the normal way gives

$$s_1 = -73 + j314 \quad \text{and} \quad s_2 = -73 - j314$$

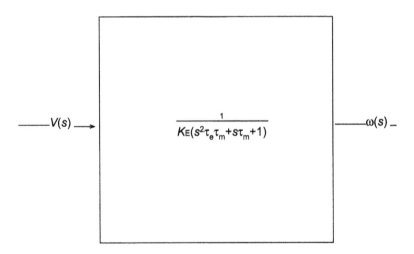

$$-V(s) \longrightarrow \boxed{\frac{1}{K_E(s^2\tau_e\tau_m + s\tau_m + 1)}} \longrightarrow \omega(s) -$$

Figure 4.10
Transfer function of motor speed relative to voltage input

Figure 4.11 shows the two poles on the s-plane. The poles are relatively close to the $j\omega$ axis which indicates that the response is oscillatory, i.e. with speed overshoot. The time constant of the decay in the oscillatory waveform is $1/73$ or

13.7 ms. Viscous damping and friction have been ignored and so the decay is the result of $i^2 R$ loss in the stator winding.

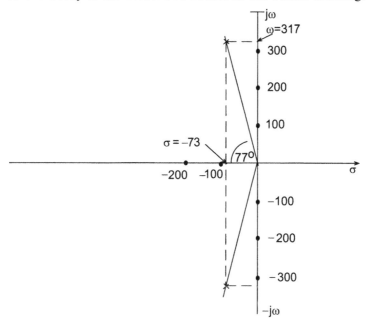

Figure 4.11
Poles of an unloaded motor with $\tau_e \approx 5\tau_m$

4.4   Variation of load inertia

Inclusion of a load inertia J_L increases the overall mechanical time constant to

$$\tau_M = (1 + J_L/J_m)\tau_m$$

The ratio of electrical to mechanical time constant falls from τ_e/τ_m to τ_e/τ_M. It is of interest to plot the locus of the poles on the s-plane as the ratio changes. The poles appear at the values of s which make the denominator of the transfer function for the motor-load unit equal to zero, that is when

$$s^2 + s\frac{1}{\tau_e} + \frac{1}{\tau_e\tau_M} = 0$$

The roots of the quadratic are

$$s = -\frac{1}{2\tau_e} \pm \sqrt{\frac{1}{4\tau_e^2} - \frac{1}{\tau_e \tau_M}}$$

When $\tau_M > 4\tau_e$, there is no sinusoidal component and the two poles lie along the horizontal axis of the s-plane. The physical interpretation of the two rates of exponential decay is that the transient response dies away relatively quickly at first, and then more slowly as time goes on.

When $\tau_M < 4\tau_e$ the response is partly exponential and partly sinusoidal. The poles appear at

$$s = \sigma \pm j\omega = -\frac{1}{2\tau_e} \pm j\sqrt{\frac{1}{\tau_e \tau_M} - \frac{1}{4\tau_e^2}}$$

For example, when $\tau_M = \tau_e$, $s = (-0.5 \pm j\sqrt{3}/2)/\tau_e$, i.e. at the 60° position.

When $\tau_M = 4\tau_e$ both poles lie at

$$s = \sigma = -\frac{1}{2\tau_e}$$

Figure 4.12 shows how the poles move as the ratio J_L/J_m rises to make the value of the overall mechanical time constant change from one-fifth to more than four times the value of the electrical time constant. For the example of the unloaded motor with the poles shown in Figure 4.11, this would mean increasing the load inertia from zero to greater than $19J_m$.

Overshoot

Figure 4.13 shows an exponential and oscillatory variation with time of the speed of an initially stationary motor-load unit, following a step input of voltage. We know that

$$\sigma = -\frac{1}{2\tau_e} \quad \text{and} \quad \omega = \sqrt{\frac{1}{\tau_e \tau_M} - \frac{1}{4\tau_e^2}}$$

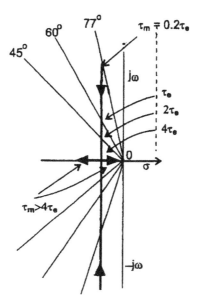

Figure 4.12
Movement of poles as load inertia is increased

The response is shown as per-unit of a final steady-state speed and has a maximum value during the first overshoot. The per-unit speed at $T/2$ is $1 + e^{-\sigma T/2}$, or

$$\frac{\omega_m}{\omega_{ss}} = 1 + e^{-\pi/2\omega\tau e}$$

When $\tau_M = \tau_e$ $\omega = \dfrac{\sqrt{3}}{2\tau_e}$ and therefore

$$\frac{\omega_m}{\omega_{ss}} = 1 + e^{-\pi/\sqrt{3}} = 1.163$$

When $\tau_M = 2\tau_e$ $\omega = \dfrac{1}{2\tau_e}$ and

$$\frac{\omega_m}{\omega_{ss}} = 1 + e^{-\pi} = 1.043$$

This means that the speed overshoot shown in Figure 4.13 increases from about 4% to 16% of a final steady-state speed as the pole position shown in Figure 4.12 is changed from

45° to 60°. As the pole angle increases, the frequency ω of the oscillatory component also increases and so the time in which ω_m reaches the zone of the first overshoot shortens. Reducing the response time in this way is, however, at the expense of increasing both the overshoot and the risk of system instability.

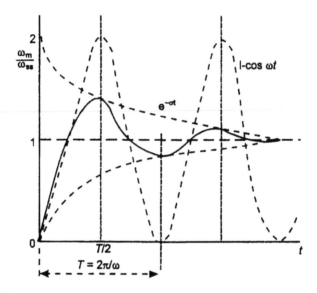

Figure 4.13
Overshoot of motor speed

System control

We have studied the transient responses which are taken into account when the motor and load are incorporated into the system as a whole. In practice, the control system is designed to eliminate the voltage step-input poles of the motor-load unit. The motor current is controlled to give the motor and load a relatively rapid change of speed, with less than 5% overshoot. The appearance of a larger than designed overshoot and associated instability is often the result of a load inertia which differs from the value used in the design of the control system. The best results are achieved when the amplifier tuning parameters include the range of inertias of the load masses likely to be driven by a particular motor [5].

4.5 Optimization

Understanding the dynamics of the motor and load is particularly important when an application involves incremental motion, where the load is required to move in discrete steps with a specific velocity profile. The steps can be in the form of an angle of rotation of a load driven through a direct or geared shaft coupling with the motor, or in the form of linear translation where the load is moved, for example, by a belt and pulley mechanism. Figure 4.14 shows a handling system used in the manufacture of filters for the automobile industry which allows three-axis translation and also rotation of the loads. The need for rapid rotation or translation often means that a load must be accelerated and decelerated back to rest over a relatively short time. Two main factors are involved in the minimization of the stator i^2R loss, and therefore of the required size and cost of a motor. We start by looking at the effect of the profile of the waveform of load velocity against time, and then go on to include the effect of the ratio between the moments of inertia of the motor and load.

Figure 4.14
Four-axis handling system. (Photo courtesy of Hauser division of Parker Hannifin)

Load velocity profiles

Figure 4.15 shows a motor connected to a rotating load through a geared reducer. The inertias J_m and J_L are assumed to include the inertias of the shaft and gear on the respective sides of the reducer. Figure 4.16 shows the general trapezoidal load velocity profile, with unequal periods of acceleration, constant speed and deceleration. Here, we are using the trapezoidal term to describe a four-sided figure with two parallel sides. By adding together the angles of rotation during the three periods and equating to the total angle θ_p, the constant speed is found to be

$$\omega_c = \frac{\theta_p}{t_p[1 - 0.5(p_1 + p_2)]}$$

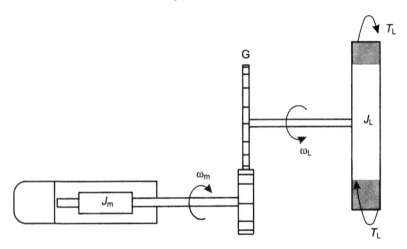

Figure 4.15
Geared drive: $\omega_m = G\omega_L$

For the symmetrical profile of Figure 4.17(a), $p_1 = p_2 = p$, and the constant speed of the load is

$$\omega_c = \frac{\theta_p}{t_p(1 - p)}$$

The rate of acceleration and deceleration of the load in the case of the symmetrical profile is

$$\frac{\mathrm{d}}{\mathrm{d}t}\omega_{\mathrm{L}} = \frac{\omega_{\mathrm{c}}}{pt_{\mathrm{p}}}$$

Combining the last two equations above and writing ω_{L} as ω_{m}/G gives the rate of acceleration and deceleration of the motor as

$$\frac{\mathrm{d}}{\mathrm{d}t}\omega_m = \frac{G\theta_{\mathrm{p}}}{t_{\mathrm{p}}^2 p(1-p)}$$

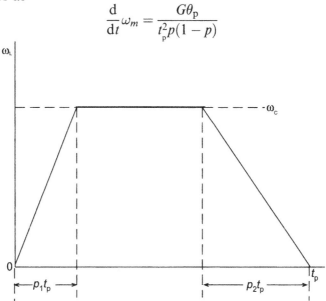

Figure 4.16
General trapezoidal, velocity profile

Stator i^2R loss

In Figure 4.15, the motor drives a rotating load through a reducer of ratio G. Torque T_{L} is assumed to be constant. If the load velocity is to follow the symmetrical profile of Figure 4.17, the motor torque required during the acceleration and deceleration periods is

$$T = K_{\mathrm{T}}i = J\frac{\mathrm{d}}{\mathrm{d}t}\omega_{\mathrm{m}} + \frac{T_{\mathrm{L}}}{G}$$

The motor current during the same periods is therefore

$$i = \frac{J}{K_{\mathrm{T}}}\frac{\mathrm{d}}{\mathrm{d}t}\omega_{\mathrm{m}} + \frac{T_{\mathrm{L}}}{GK_{\mathrm{T}}}$$

or

$$i = I_1 + I_2$$

where I_1 is the component of motor current required for the constant acceleration or deceleration of the load and I_2 is the component which provides a constant output torque. Figure 4.17(b) shows the current waveform (rms in the case of the sinusoidal motor). Note that I_1 is negative during deceleration. The total energy in joules produced in the form of heat by the $i^2 R$ loss is

$$\epsilon = [p(I_2 + I_1)^2 + (1 - 2p)I_2^2 + p(I_2 - I_1)^2]Rt_p \tag{J}$$

$$= (2pI_1^2 + I_2^2)Rt_p$$

giving

$$\epsilon = \frac{Rt_p}{K_T^2}\left[2p\left(J\frac{d}{dt}\omega_m\right)^2 + \left(\frac{T_L}{G}\right)^2\right]$$

Replacing $\frac{d}{dt}\omega_m$ with the form already found in terms of G, θ_p, t_p and p gives the stator heating energy as

$$\epsilon = \frac{Rt_p}{K_T^2}\left[c_p\left(\frac{JG\theta_p}{t_p^2}\right)^2 + \left(\frac{T_L}{G}\right)^2\right]$$

where the profile constant is $c_p = \dfrac{2}{p(1-p)^2}$.

For a given motor, load, and reducer ratio G, the stator heating is at a minimum when

$$\frac{d}{dp}\epsilon = 0, \quad \text{i.e. when} \quad \frac{d}{dp}\frac{2}{p(1-p)^2} = 0$$

giving

$$p = \frac{1}{3}$$

The most efficient symmetrical profile is therefore equally distributed, as shown in Figure 4.18. When $p = 1/3$, the profile constant is

$$c_p = 13.5$$

The last expression for ϵ above can be shown to apply when the load velocity follows the general trapezoidal profile of Figure 4.16. The profile constant becomes

$$c_p = \frac{p_1^{-1} + p_2^{-1}}{[1 - 0.5(p_1 + p_2)]^2}$$

The constant has a value greater than 13.5 at all relevant values of p_1 and p_2 other than $p_1 = p_2 = \frac{1}{3}$. Stator heating is therefore at a minimum when the trapezoidal profile of load velocity is equally distributed.

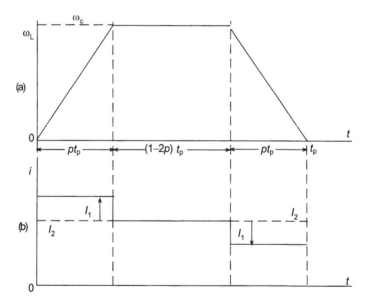

Figure 4.17
Motor current for the symmetrical, load velocity profile

The second term of the last expression above for ϵ does not depend on c_p, and profile optimization is unnecessary when the load torque requires most of the $K_T I$ product. Optimization of the velocity profile can be of benefit if the acceleration and deceleration of the load mass is responsible for a high proportion of the total stator $i^2 R$ loss. Such cases are likely to arise when a substantial load mass is moved rapidly and repeatedly.

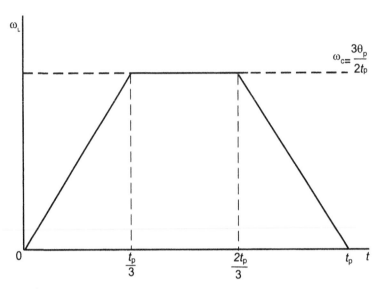

Figure 4.18
Optimum trapezoidal profile for incremental motion

The inertia match for the geared drive

When the speed of a load is changed many times per second, backlash in the transmission components can obviously cause problems. Gear reducers used for incremental motion must be of very high quality and are relatively expensive. A planetary reducer with a maximum backlash of 2 to 3 arcminutes may have the same order of price as the drive motor itself. We have already seen how the load velocity profile affects the motor losses, and now follow on by including the effect of the ratio between the inertias of the motor and load masses.

The inertial load

In Figure 4.15 the load is purely inertial when load torque T_L is zero. Ignoring reducer losses, motor friction and viscous damping, the motor torque is given by

$$T = K_T i = J\frac{\mathrm{d}}{\mathrm{d}t}\omega_m$$

where the sum of the motor inertia and the equivalent load inertia reflected at the motor side of the reducer is

$$J = J_m + \frac{J_L}{G^2}$$

Writing ω_m as $G\omega_L$, the motor current is therefore

$$i = \frac{JG}{K_T} \frac{d}{dt} \omega_L$$

The energy dissipated in the form of heat in the motor stator over the time dt is

$$d\epsilon = i^2 R dt$$

Suppose that the load is to move a complete step in a time interval of t_p over *any* load velocity profile. Using the last three equations above and integrating gives the energy dissipated in the stator winding as

$$\epsilon = \frac{RJ^2 G^2}{K_T^2} \int_0^{t_p} \left(\frac{d}{dt} \omega_L \right)^2 dt$$

Varying the reducer ratio minimizes the stator heating energy when

$$\frac{d}{d(G)^2} \epsilon = 0$$

The integration term in the above expression for ϵ depends on the velocity profile of the load during time t_p, but the profile itself does not depend on the gear ratio. Stator heating is therefore at a minimum for any particular profile when

$$\frac{d}{d(G)^2} G^2 (J_m + \frac{J_L}{G^2})^2 = 0, \quad \text{i.e. when} \quad G_0 = \sqrt{\frac{J_L}{J_m}}$$

G_0 is the gear ratio which minimizes the motor stator heating during incremental rotation of a purely inertial load, for any velocity profile.

Note that although G_0 is independent of the velocity profile, the same is not true of the heating energy ϵ. However, for any particular combination of motor, inertial load and velocity profile, the energy is a function of the gear ratio alone. For reducer ratios G and G_0, the energies are

$$\epsilon = \gamma G^2 (J_m + \frac{J_L}{G^2})^2 \quad \text{and} \quad \epsilon_0 = \gamma G_0^2 (J_m + \frac{J_L}{G_0^2})^2$$

where γ is a constant. Dividing ϵ by ϵ_0 and rearranging gives the extra heating factor for a non-optimum gear ratio as

$$\frac{\epsilon}{\epsilon_0} = 0.25 \left(\frac{G}{G_0} + \frac{G_0}{G} \right)^2$$

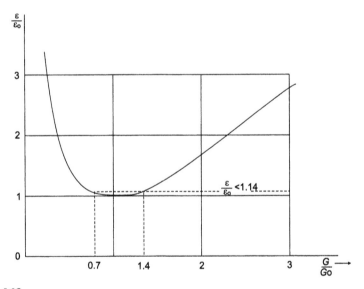

Figure 4.19
Increase in stator heating when $G \neq G_0 (T_L = 0)$

Ref. [4] develops the above expression and plots the energy ratio as the gear ratio falls in comparison to the optimum value. The curve has the same shape when it is plotted for rising values of gear ratio in comparison to the optimum. Figure 4.19 shows how the $i^2 R$ loss increases as the gear ratio deviates from the ideal. As a rule of thumb:

$$\frac{\epsilon}{\epsilon_0} < 1.14 \quad \text{when} \quad 0.7 < \frac{G}{G_0} < 1.4$$

Note that the stator heating doubles when the gear ratio is more than twice or less than half the optimum value.

Effect of load torque on the optimum gear ratio

Let us now return to the case where the load is due partly to the inertias of the motor rotor and the load, and partly to a constant torque which is delivered over a specific velocity profile. The stator heating energy when the load velocity follows the general trapezoidal profile of Figure 4.16 has been shown to be

$$\epsilon = \frac{R t_p}{K_T^2} \left[c_p \left(\frac{JG\theta_p}{t_p^2} \right)^2 + \left(\frac{T_L}{G} \right)^2 \right]$$

Writing J in terms of J_L, J_m and G as before and rearranging gives the heating energy as

$$\epsilon = c_p R \frac{\theta_p^2 J_L^2}{K_T^2 t_p^3} \left[G^2 \left(\frac{J_m}{J_L} + \frac{1}{G^2} \right)^2 + \frac{\Delta_1}{G^2} \right]$$

where

$$\Delta_1 = \frac{1}{c_p} \left(\frac{T_L t_p^2}{\theta_p J_L} \right)^2$$

Differentiating the last expression for ϵ above with respect to G^2 and equating to zero gives the gear ratio for minimum stator heating as

$$G_\Delta = G_0 (1 + \Delta_1)^{0.25}$$

G_Δ is the gear ratio which minimizes stator heating when the load consists of an inertial mass and an output torque, for any trapezoidal velocity profile.

We have optimized the ratio of a reducer for any trapezoidal velocity profile, and we found earlier that the most efficient shape is equally distributed. The trapezoidal profile has the advantage of ease of control by the drive. It is not, however, the most efficient velocity profile in general. The lowest losses of all occur when the profile of load velocity against time is parabolic [4].

Example 4.1

A sinusoidal motor is connected to a load through a geared reducer. The load is to be rotated incrementally, using an equally distributed, trapezoidal velocity profile. The motor details are given in Table 4.1. The system constants are as follows:

$$J_m = 0.00022 \, kgm^2$$
$$\theta_p = 2 \, rad$$
$$t_p = 0.06 \, s$$
$$c_p = 13.5$$
$$T_L = 10 \, Nm$$
$$J_L = 0.0022 \, kgm^2$$

If there were no opposing load torque, Δ_1 would equal zero and

$$G_0 = \sqrt{\frac{J_L}{J_m}} = 3.2$$

For the constant opposing torque of 10 Nm,

$$\Delta_1 = \frac{1}{c_p} \left(\frac{T_L t_p^2}{\theta_p J_L} \right)^2 = 5.0$$

giving

$$G_\Delta = 3.2(1 + 5.0)^{0.25} = 5.0$$

The total stator heating energy is formed from two components. One arises from the effort of rotating the load mass from one stationary position to another, increasing as already shown in Figure 4.19 as G is changed from G_0. The second component is generated when the load mass is subjected to the opposing torque T_L. The motor output torque required (for a constant T_L) falls as G rises, together with the associated stator loss. When G is changed from 3.2 to 5.0 in the present example, the extra loss due to the mismatch of motor to load inertia is compensated by the fall in the loss associated with the supply of T_L.

One way to study the relative importance of the two components of the stator heating energy is to calculate the values as the step time t_p is varied, all other parameters (including G) remaining constant. Figure 4.20 shows the variation in the two energies in the present example when $G = 5.0$. The motor torque required for load translation rises as t_p falls, generating rapidly rising i^2R losses below $t_p = 100$ ms. On the other hand, the stator heating due to the supply of T_L becomes dominant as t_p is increased. A relatively large increase in G would then be needed if the overall losses are to be minimized at the higher values of t_p.

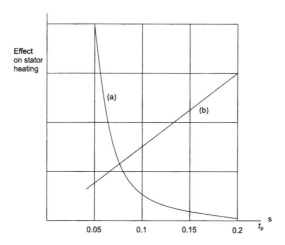

Figure 4.20
Relative stator heating due to (a) load translation and (b) load output torque

Limitations to minimization of the i^2R loss

Minimization of the i^2R loss is desirable when a significant part of the stator heating is due to the effect of load and motor inertia on the motor torque requirement. Minimization may still be worthwhile when the output load torque forms a relatively high proportion of the total torque, but the optimum gear ratio (and therefore the motor speed) rises with the output torque. A non-optimum solution is often imposed by practical limitations on the speed of the motor and reducer input, and by reducer losses at the higher speeds.

The belt and pulley drive

Belt drives are often used in pick-and-place robotics, for example in the loading of electronic components onto printed circuit boards. Figure 4.21 shows a load attached to a belt and pulley which may be driven directly by the motor, or through a reducer of ratio G. The reducer may be in the form of a gear train, or another belt and pulley system. Mass m is the total mass of the load and the load conveyor belt. If present, force F is assumed constant over the step of length x.

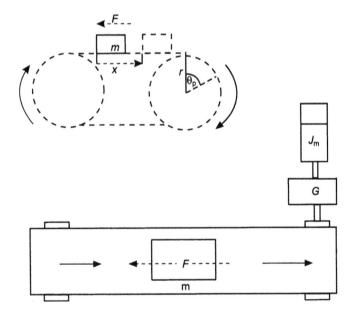

Figure 4.21
Belt and pulley drive

The system can be optimized at the reducer or at the drive pulley. If optimization is to take place at the reducer, the optimum ratio is

$$G_\Delta = \sqrt{\frac{J_L}{J_m}}\,(1+\Delta_1)^{0.25} \quad \text{where} \quad \Delta_1 = \frac{1}{c_p}\left(\frac{T_L t_p^2}{\theta_p J_L}\right)^2$$

From the diagram, $\theta_p = x/r$

$$T_L = Fr$$
$$J_L = mr^2$$

The optimum ratio of a reducer for the belt and pulley drive is therefore

$$G_\Delta = r\sqrt{\frac{m}{J_m}}\,(1+\Delta_2)^{0.25} \quad \text{where} \quad \Delta_2 = \frac{1}{c_p}\left(\frac{Ft_p^2}{xm}\right)^2$$

When the use of an existing reducer or a direct drive is convenient, the belt and pulley mechanism may be optimized by selecting the correct radius of the drive pulley. Take first the case where there is little or no opposing force F, making $\Delta_2 = 0$. Writing r_0 as the pulley radius which makes the ratio G' of the existing reducer equal to the optimum value gives

$$G' = r_0\sqrt{\frac{m}{J_m}}$$

The optimum pulley radius when a reducer of nominal ratio G' is already in place is therefore

$$r_0 = G'\sqrt{\frac{J_m}{m}}$$

where $G' = 1$ when the motor is coupled directly to the pulley. r_0 is the optimum radius of the belt drive pulley when there is no opposing load force, for any velocity profile.

Working in the same way for the case where $F \neq 0$ gives

$$r_\Delta = G'\sqrt{\frac{J_m}{m\sqrt{1+\Delta_2}}}$$

r_Δ is the optimum pulley radius when the load includes an opposing force, for any trapezoidal velocity profile.

The inertias of the drive pulley and reducer have so far been ignored. Where the motor shaft is connected directly to the drive pulley and optimization takes place according to the drive pulley radius, the inertia of the drive pulley should be added to the motor inertia and a correction made to the optimum pulley radius. Where optimization takes place according to the ratio G_Δ of a reducer, G_Δ should be corrected by adding to mass m the mass which would have the same inertia as that of the reducer output plus drive pulley, if distributed at the drive pulley circumference. The inertia of the reducer input should be added to J_m.

Example 4.2

A load is to be belt driven and moved incrementally. The force opposing the load is negligible. The following are already available:

(a) The motor of the previous example.
(b) A reducer of ratio $G' = 3.5$.
(c) A belt and pulley drive.

The system constants are:
$$J_m = 0.00022\,kgm^2$$
Mass of belt and load $m = 1.0\,kg$
Existing drive pulley radius $r = 0.03\,m$

The load force is negligible in this case, and so the load velocity profile, the step time t_p and the step distance x have no effect on the optimization. The optimum pulley radius is

$$r_0 = G'\sqrt{\frac{J_m}{m}}$$

Two or three mechanical configurations are possible:

1. No reducer and a drive pulley radius of

$$r_0 = \sqrt{\frac{0.00022}{1.0}} = 0.015\,m$$

This radius is close to the minimum for toothed drive belts.

2. Use the existing reducer and a new drive pulley with a radius of

$$r_0 = 3.5 \sqrt{\frac{0.00022}{1.0}} = 0.052 \text{ m}$$

3. Use the existing drive pulley and change the ratio of the reducer to

$$G_0 = 0.03 \sqrt{\frac{1.0}{0.00022}} = 2.0$$

Finally, a correction should be made by taking into account the inertias of the prospective pulley or reducer.

The ball screw and lead screw drives

Figure 4.22 shows the principle of the ball screw and lead screw mechanisms, the screw may be driven by the motor, or through a belt and pulley or geared reducer. Screw mechanisms can provide high forces at the load, and also allow accurate load positioning. Figure 4.23 shows a screw drive being used to control the vertical position of a two-axis handling machine used in the internal coating of glass medical bottles. In this machine, the horizontal position is controlled by a belt and pulley drive.

The primary and secondary threads of a ball screw drive are connected through ball bearings. This results in low backlash and a resolution at the load which can be better than 1 micron. The transmission of energy is very efficient, the losses being low enough to allow the device to be 'backdriven' from the load side. The disadvantage of this is that the ball screw may be backdriven by the load if the motor torque is lost, say through a power failure. Consequently, the ball screw driving motor must be fitted with a brake.

In the case of the lead screw, the threads are in direct contact. This results in a transmission which is inefficient in comparison to the ball screw, but which does not backdrive following the loss of motor torque and is relatively inexpensive. The lead screw inevitably has the backlash associated with screw threads.

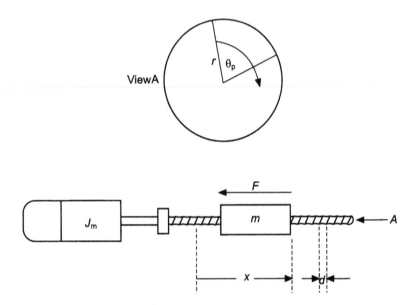

Figure 4.22
The screw drive

The load mass m travels distance x in time t_p, for both the belt and pulley drive of Figure 4.21 and the screw drive of Figure 4.22. The load inertia is mr^2 for both (in the case of the screw, imagine that the load is travelling around a stationary screw). The screw driven load travels distance d for one revolution of the screw, and so the screw drive has a 'gear ratio' of $2\pi r/d$. The inertia of the load mass reflected on the motor side of the screw threads is therefore

$$J_1 = mr^2 \left(\frac{d}{2\pi r} \right)^2 = \frac{md^2}{4\pi^2}$$

The reflected inertia of the load mass should be matched with the inertia which already exists on the motor side of the screw threads, and so

$$J_1 = J_m + J_{sw}$$

where J_{sw} is the inertia of the screw. Combining the last two equations and including the effect of force F gives the optimum screw pitch as

$$d_\Delta = 2\pi\sqrt{\frac{J_m + J_{sw}}{m\sqrt{1 + \Delta_2}}} \quad \text{where} \quad \Delta_2 = \frac{1}{c_p}\left(\frac{Ft_p^2}{xm}\right)^2$$

d_Δ is the optimum screw pitch when the load follows any trapezoidal velocity profile and is subject to an opposing force.

Figure 4.23
Two-axis, pick-and-place handling machine. (Photo courtesy of Hauser division of Parker Hannifin)

Example 4.3

The sinusoidal motor in Table 4.1 is to be used to drive a ball screw. The load velocity is to follow the profile shown in Figure 4.24. The system constants are

$$J_m = 0.00022 \, kgm^2$$
$$J_{sw} = 0.00003 \, kgm^2$$
$$x = 0.025 \, m$$
$$t_p = 0.120 \, s$$
$$F = 1000 \, N$$
$$m = 5 \, kg$$

The profile constant is $c_p = \dfrac{p_1^{-1} + p_2^{-1}}{[1 - 0.5(p_1 + p_2)]^2}$.

In this case $p_1 = 20/120$ and $p_2 = 60/120$, giving $c_p = 18$. The load force factor is

$$\Delta_2 = \frac{1}{18} \left(\frac{1000 \times 0.12^2}{0.025 \times 5} \right)^2 = 737$$

The optimum screw pitch is

$$d_\Delta = 2\pi \sqrt{\frac{0.00022 + 0.00003}{5\sqrt{1 + 737}}} = 8.5 \, \text{mm}$$

In this example, the optimum pitch is dictated mainly by the effect of the load force. In the hypothetical case of zero load force, the optimum screw pitch given by the calculation would be $d_0 = 44$ mm!

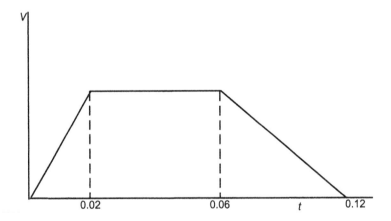

Figure 4.24
Load velocity profile for example 4.3

4.6 Torsional resonance

In Sections 4.4 and 4.5 the mechanical connection between the motor and load is assumed to be inelastic, leaving the increase in system inertia as the only mechanical effect of an added load. In practice some flexibility in the connection is unavoidable, and an error may develop in the position of the load relative to that of the hub of the rotor of the motor as torsional forces come into play. When the error becomes oscillatory, the condition is known as *torsional resonance*. The problem can also arise in the section of shaft between the hub and the sensor, but we will assume throughout that any such effects have been eliminated through the design of the motor and sensor. Under these circumstances, the error between the sensor and the load can be assumed to be the same as the error between the hub and the load.

Figure 4.25 shows a rotor of inertia J_m connected to a rotating load through a shaft which is subject to twist. Following the approach used in Section 4.4 for the totally rigid shaft shows that the poles of the transfer function of motor speed response are given by

$$J_L[s^4 \tau_e \tau_m + s^3 \tau_m + s^2] + C^{-1}[s^2 \tau_e \tau_M + s\tau_M + 1] = 0 \qquad (1)$$

where $\tau_M = \tau_m(J_m + J_L)/J_m$. If $J_L = 0$, $\tau_M = \tau_m$ and the equation reduces to the form already derived for an unloaded motor:

$$s^2 \tau_e \tau_m + s\tau_m + 1 = 0$$

C is the *compliance* of the mechanical connection, or error factor for the angle between the motor position sensor and the driven load, normally expressed in microradians/Nm. The compliance varies widely according to the length and diameter of the drive shaft and the types of transmission between the motor and the load. Typical values are in the range 10–100 μrad/Nm.

Solution of expression (1) gives the theoretical locations of the poles for the case where the damping effects of eddy currents and friction are ignored. There are two pairs of poles, one pair at low frequency and the other at the frequency of potential torsional resonance.

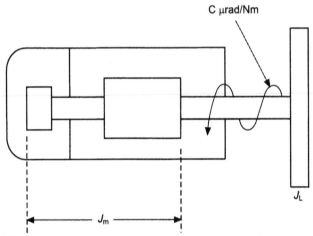

Figure 4.25
Shaft compliance

Effect of compliance at a fixed load inertia

The resonant frequency is affected by the ratio of the motor to load inertias, and also by the value of compliance. We start by looking at the way the resonant frequency varies as the compliance is changed, for a case where the inertias are approximately equal.

Example 4.4

A rotating load is connected to the shaft of a brushless servomotor. The motor and load inertias are approximately equal. Torsional resonance frequency values are required for a wide range of shaft compliance. The system constants are

$$J_m = 0.00215\,kgm^2$$
$$J_L = 0.00200\,kgm^2$$

$$\tau_e = 5.0\,ms$$
$$\tau_m = 2.6\,ms$$
$$\tau_M = 5.0\,ms$$

Inserting the numerical values in equation (1) above and dividing through by $J_L \tau_e \tau_m$ gives

$$s^4 + 200s^3 + (962C^{-1} + 77 \times 10^3)s^2$$
$$+ 192 \times 10^3 C^{-1} s + 38 \times 10^6 C^{-1} = 0$$

Figure 4.26 shows how the poles move as C is varied from infinity to 10 μrad/Nm. The physical interpretation of infinite compliance is of course that the load is disconnected from the motor, at which point the last expression is reduced to

$$s^4 + 200s^3 + 77 \times 10^3 s^2 = 0$$

The motor and disconnected load therefore have four poles, two showing the normal response (already dealt with in Section 4.3) of an unloaded motor to a step voltage input. The other two poles remain at the origin as long as the load stays unconnected. The arrows show the shift in position of the four poles as the compliance is reduced from infinity, or in other words as the stiffness of the transmission is increased from zero.

As the compliance is reduced, the motor-load poles move from the position (at the origin) for a disconnected transmission towards the positions P_1, P_2 for a totally rigid connection. The poles at positions P_3, P_4 for the normal response of the unloaded motor rise in frequency but become increasingly oscillatory as the compliance falls, taking up relatively undamped positions close to the boundary between stable and unstable operation of the system. In practice it is found that the lower the frequency of such lightly damped responses, the more likely it becomes for the frequency to be excited by the system in general and for system instability to occur.

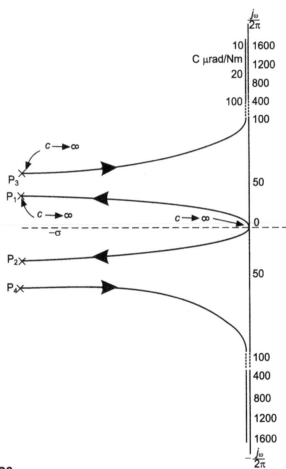

Figure 4.26
Pole loci as compliance is reduced

Resonant frequency predictions and tests

The low frequency poles in Figure 4.26 are relatively well damped and are in any case eliminated through the design of the control system. The other pair travel towards infinity as the compliance falls towards zero. In the present case the resonant frequency is predicted to lie between approximately 1550 and 1100 Hz for compliance values from 10 to 20 μrad/Nm.

The motor and load of Example 4.4 were connected together. The compliance of the length of shaft between the front end of the hub of the rotor and the load was calculated to be $C = 14.4$ μrad/Nm. The resonant frequency was excited by striking the load, and measured by recording the stator emf produced by the resulting oscillations of the rotor. The result is shown in Figure 4.27(a). The low frequency envelope is due to the slow rotation of the rotor after the shaft has been struck. The resonant frequency is close to 1305 Hz, and this compares well with the value of 1300 Hz predicted by equation (1) at the compliance of 14.4 μrad/Nm.

As the resonant frequency rises, it becomes less likely to be excited through a well-designed drive system. The resonant frequency rises as the compliance falls, and so the main conclusion is that the compliance should be as low as possible for maximum system stability.

Damping

Expression (1) automatically includes the damping effects of the i^2R loss generated in the stator as the rotor oscillates, but these are negligible. The pole loci in Figure 4.26 do not take account of damping due to frictional and eddy current losses. When expression (1) is modified to include the viscous damping due to eddy currents, the effect is predicted to be insignificant in the test motor. The test results do not include the effects of any i^2R loss in the stator as measurements must be made with the winding on open-circuit. Damping of the motor under such test conditions is therefore the result of eddy current loss and losses at the bearings, with the bearing loss likely to be the greater part. The time constant of the decay in Figure 4.27(a) is approximately 33 ms.

The time constant affecting the rate of decay of the oscillations is seen to be high, when the load is mainly inertial. The rate of decay is of course increased in practice when the driven load is subject to friction, and also when damping appears in a transmission mechanism such as a belt and pulley drive. As friction and damping increase, the resonant frequency becomes less likely to be excited by the system [6], [7].

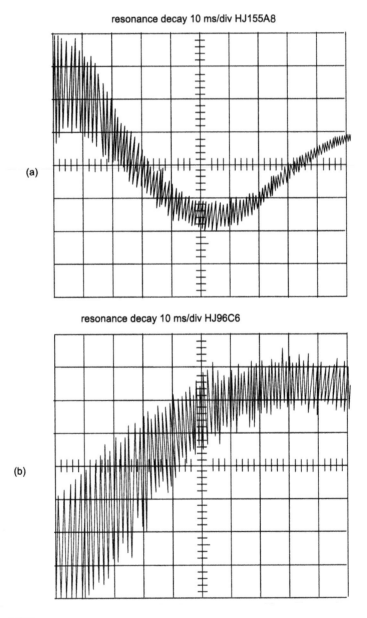

Figure 4.27
Open-circuit stator emf during torsional resonance

The effect of load inertia

In Example 4.4, the variation of the motor-load pole positions over a range of shaft compliance was plotted for the case where the motor and load inertias are approximately equal. In many applications a close match of inertias for the purpose of minimization of the i^2R loss may be unnecessary or impracticable, and so in practice the load inertia may be several times that of the motor. In order to study the effect of load inertia on the resonant frequency, equation (1) may be rearranged as

$$J_r J_m [s^4 \tau_e \tau_m + s^3 \tau_m + s^2] + C^{-1}[s^2 \tau_e \tau_m (1 + J_r) \\ + s \tau_m (1 + J_r) + 1] = 0 \qquad (2)$$

where $J_r = J_L/J_m$. We can now forecast the resonant frequencies as J_r varies, for a fixed value of compliance.

Example 4.5

Values of the resonant frequency of a motor-load combination are required for load to motor inertia ratios from 1.0 to 10. The motor constants are

$$J_m = 0.000315 \, kgm^2$$
$$\tau_e = 2.3 \, ms$$
$$\tau_m = 2.8 \, ms$$
$$C = 67.5 \, \mu rad/Nm$$

Figure 4.28 shows the predicted variation in the resonant frequency when the above values are used in expression (2). As J_r varies from 1 to 10, the resonant frequency falls from 1540 to 1145 Hz. Note that the position on the curve for matched inertias is not in any way a special point, and that the matched inertia case has no significance as far as resonance is concerned. In this particular example, the frequency is predicted to fall by 25% as the inertia ratio rises from 1 to 10. The fall is rapid at first and then levels off, and most occurs for a ratio of only 4:1.

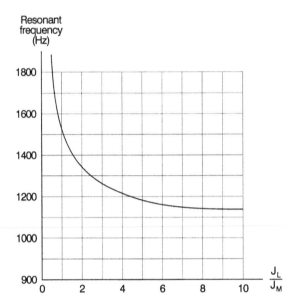

Figure 4.28
Fall in resonant frequency with increasing load inertia

Predictions and tests

A rotating load with an inertia of 0.002 kgm^2 was fitted to the shaft of the motor of Example 4.5. This gave a ratio of load to motor inertia of

$$J_r = 6.35$$

The resonant frequency and decay were measured by the method used for the previous example. The results are shown in Figure 4.27(b). The resonant frequency is 1167 Hz, which again compares well with the predicted value of 1175 Hz. Damping of the oscillations is again low, the time constant of decay being approximately 20 ms.

Compliance and inertia in practice

We have looked at the effects of the compliance of the motor shaft. In practice there may be several other points in the transmission mechanism which either add to the compliance problem, or help by providing damping. In the example of the ball screw drive, additional compliance occurs at the coupling between the motor shaft and the screw input shaft, and also along the complete length of the input shaft and screw. The screw itself makes a further contribution to the difficulties by having a compliance which increases as the load moves away from the motor. Damping is added by losses at the screw input bearing and the point of screw contact with the load. Such systems are difficult to analyse by the method used above for the simple case where the load is connected directly to the end of the motor shaft, and modelling using electrical circuit analogues can be a better approach [8].

In general the resonant frequency falls as the compliance increases, and also as the moment of inertia of the load increases in relation to that of the motor. Assuming as much as possible has been done to reduce the compliance, the next step is to reduce the inertia ratio. This can be done by lowering the effective value of the load inertia by means of a reducer, but this method may be prevented by economic and practical considerations. Where this is the case, the best way forward may be to reduce the inertia ratio by increasing the inertia at the motor end of the mechanical drive link. An increase in motor inertia can be made in two ways. One method has been to introduce the required extra inertia by fitting an oversize motor, but this is normally an expensive solution. The other way is to fit a motor of the required size, torque rating and price which has been designed around an increased rotor inertia. Such motors are available over a wide range of servomotor ratings, and offer higher stability for systems with a relatively high load inertia which cannot be reduced.

CHAPTER 5

MOTOR RATING AND SELECTION

5.1 Introduction

This chapter explains how a motor should be rated for use on steady-state, intermittent or incremental duty. For steady-state applications where the motor output is defined in terms of a constant and continuous speed and torque, motor rating and selection from a range of motor specifications is normally straightforward. If the motor torque is required intermittently, selection is guided by specifications which give torque and speed limits for intermittent duty in general. Care is needed here in assessing the likely effects of the intermittent operation of the motor on the peak rise in motor temperature.

In incremental motion the requirements are not usually defined only in terms of motor output torque and speed. The other loading conditions normally include load inertia and angle of rotation, or load mass and distance covered, plus the load velocity profile over the period of movement. The required motor speed and torque profiles may be calculated for individual motors on a trial and error basis, and the experienced designer may often select the most suitable motor without too much iteration. To assist in the process, a selection method is given in this chapter which indicates the most suitable motor for given incremental loading conditions.

Motor suppliers normally provide all the data which the user needs to assess the thermal performance of a motor, but the values of the constants should never be assumed to apply exactly to a particular motor. Calculations which give the final temperature for the most likely 'worst case' normally assume the stator resistance to be 10% higher, and the torque constant 10% lower than the nominal values given in the motor specification. The numerical examples in this chapter show calculations of nominal torques and temperatures using nominal constants, and also give the results at the extreme tolerance of the constants.

5.2 Motor heating

Heat is generated in the brushless motor as a result of the i^2R loss in the stator winding and iron losses in the stator and rotor. In addition, some heat arises from the friction between the bearing seal and the rotor shaft. Rotor windage loss is normally very low. Iron losses are produced by induced eddy currents in the stator and rotor, and are partly a function of the PWM frequency of the stator winding current. Motors which exhibit rotor heating do not appear to have lower losses when fitted with laminated rather than solid hubs. This is because the eddy currents favour the magnets as the closest part of the rotor to the stator winding. Laminated hubs do of course have advantages as far as manufacturing costs are concerned.

Motor temperature rises as losses are generated, and the thermal resistance R_{th} quoted by the manufacturer gives a temperature rise in degrees centigrade per watt of total power loss. However, the final steady-state temperature differs at the various parts of the motor structure; for example, the magnets may become hotter than the stator core. The figure for R_{th} must therefore be high enough to allow any part of the motor to remain below the maximum temperature, normally 150°C.

Soac curves

The thermal characteristic of a brushless motor is usually given as the boundary of the Safe Operation Area for Continuous operation, which takes the form shown in Figure 5.1. The curve shows the maximum continuous torque rating at all speeds from zero to maximum in the way already described for the brushed motor in Chapter 1.

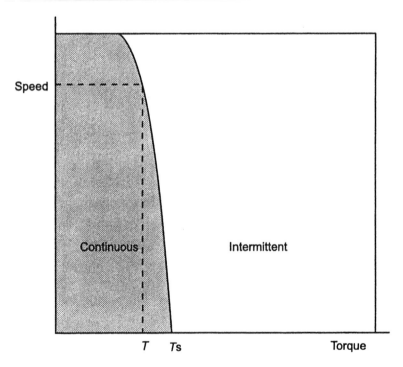

Figure 5.1
The Soac thermal boundary

The Soac curve is drawn from the results of practical tests. The motor is loaded and the current gradually adjusted until the above-ambient steady-state temperature at any part of the motor reaches 110°C. This means that at an ambient temperature of 40°C, no part of the motor exceeds the maximum allowable figure of 150°C. The speed and current are measured, and the torque is calculated by multiplying the current by the torque constant of the motor. (K_T is independent of motor speed.) At zero speed the rated stall torque is

$$T_s = K_T I_s$$

where I_s is the maximum continuous current at zero speed, when the heating is entirely the result of the i^2R loss.

Speed-sensitive loss

If the motor could operate without generating a speed-sensitive loss, the Soac characteristic would rise vertically from T_s. The total losses and motor temperature are constant and maximum at all points on the curve. The i^2R loss at speed is therefore lower than that at zero speed by the amount of the speed-sensitive loss P_{sp}. At all points on the curve the total losses are therefore

$$I^2R + P_{sp} = I_s^2R'$$

giving

$$P_{sp} = (I_s^2 - I^2)R'$$

Resistance R' is the stator resistance at the maximum motor temperature. Motor specifications normally give the 'cold' stator resistance R at 25°C, and so at 150°C the resistance is

$$R' = R[1 + 0.00385(150 - 25)] = 1.48\,R$$

where the figure of 0.00385 is the temperature coefficient of resistance of copper for a temperature rise above 25°C. At any speed, the speed-sensitive loss is therefore given by

$$P_{sp} = 1.48\,\frac{R}{K_T^2}\,(T_s^2 - T_{soac}^2)$$

where T_{soac} is the rated torque given by the Soac curve. The speed-sensitive loss may amount to 25–35% of the total power loss when the brushless motor runs on load at the centre of the speed range. A restriction on the average operation period of the motor may be imposed at mid to high speeds, where a substantial speed-sensitive loss is possible.

Confusion sometimes arises when the horizontal axis of the Soac plane is treated as a torque which is related solely to motor speed. This misinterpretation allows an erroneous 'torque loss' to be calculated by subtracting the torque at a point on the curve from the torque at the point of stall. It should be remembered that the horizontal axis gives the torque available at various speeds when the motor temperature is at its maximum limit. Differences between the *squares* of the rated torques shown along the axis are used when the speed-sensitive loss is calculated, as already described.

Heatsinks and blowers

The Soac curve is drawn from the results of rating tests in which the motor is attached to a heat-absorbing frame. When the motor is installed in its working environment, its front flange is usually bolted directly to the frame of the driven machinery, the frame providing a heatsink. If the frame has been thermally isolated from the motor mounting point in some way, a heatsink should be included between the motor flange and the frame.

There are two situations where the cost of a blower may be justified at the planning stage of an installation. One is where a combination of a very low motor inertia and a high torque is required. Low inertia limits the choice of motor to the small end of the range, where output torque ratings are also relatively low. Forced cooling must then be used to increase the rating. The other case where a blower is sometimes used is at the other end of the motor range. If the very largest machine available cannot provide the output required, its rating may be increased with a blower. Between these two extremes, blowers are not usually part of the best solution to a particular load specification and should be avoided if possible. The cost of a blower and motor is normally of the same order as the cost of the next larger motor. Any saving in cost is clearly negligible in comparison to the cost of failure of the blower, followed by the motor.

5.3 Steady-state rating

The rating of a brushless motor in terms of its continuous, constant torque output may be assessed in the way described in Chapter 1 for the brushed machine. The maximum continuous stall current I_s is normally given in the data sheet, and the Soac curve is always available. If the form factor of the current supplied to a trapezoidal motor or to a brushed DC motor is other than unity, the same considerations apply to both.

Example 5.1

A motor is to run continuously. Estimate the speed-sensitive loss using the Soac diagram of Figure 5.2 (motor M06). Estimate the stall torques, and the maximum output torques at a speed of 4500 rpm, available from the trapezoidal and sinusoidal forms of the motor. The form factor of the current supplied to the trapezoidal motor is 1.1. The motor constants are

$$K_T(trap) = 0.42 \, Nm/A$$
$$K_T(sin) = 0.51 \, Nm/A$$
$$R_{LL} = 1.29 \, \Omega$$

Speed-sensitive loss

Referring to Figure 5.2 (M06) and using the trapezoidal constants gives the estimate of the loss at 4500 rpm as

$$P_{sp} = 1.48 \frac{R}{K_T^2} (T_s^2 - T_{soac}^2) = 1.48 \frac{1.29}{0.42^2} (3.7^2 - 3.2^2) = 37 \, W$$

The figure of 37 W applies to both motors, remembering that the thermally effective resistance for the sinusoidal form is $R = 1.5\,R_{LL}$.

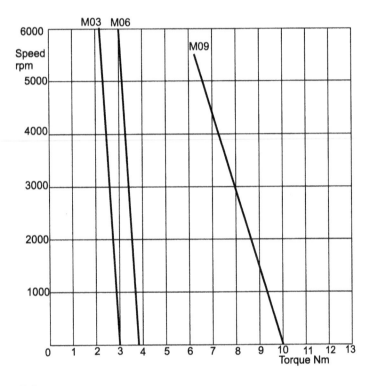

Figure 5.2
Soac curves

Torques

Trapezoidal motor

The maximum continuous output torque at 4500 rpm is 3.2 Nm when the motor current has a form factor of unity. When the form factor is 1.1, the torque must be reduced by 1/1.1 so that the rms value of the input current remains the same. The maximum average torque available at 4500 rpm is then

$$T = \frac{3.2}{1.1} = 2.9\,\text{Nm}$$

and the average, continuous stall torque becomes

$$T_s = \frac{3.7}{1.1} = 3.3 \, \text{Nm}$$

Sinusoidal motor

If the input current is purely sinusoidal, the maximum continuous torque at 4500 rpm is $T_{\text{soac}} = 3.2 \, \text{Nm}$.

5.4 Intermittent torque

Intermittent operations may be broadly divided into two types. This section deals with the simple torque profile, where there are intervals between the application of a constant motor torque. Cases where the load moves incrementally over a specific velocity–time profile are left until Section 5.5.

Motor temperature

The main effects of intermittent operation on the temperature of the brushed motor were dealt with in Chapter 1. It was seen that the intermittent rating of the brushed motor is affected by the presence of a temperature ripple, which is most pronounced at the rotor winding. The same effects occur at the stator winding of the brushless motor. Figure 5.3 shows an intermittent output torque, applied every t' seconds over a time of t_p. The remainder of the diagram shows the associated power loss and the steady-state temperature of the winding after the motor has been running on the intermittent cycle for some time. The diagram is similar to that of Figure 1.15, and shows above-ambient winding temperatures of Θ_{min}, Θ_{av} and Θ_{pk}. The average, above-ambient temperature is

$$\Theta_{\text{av}} = R_{\text{th}} P_{\text{loss(av)}}$$

where $P_{\text{loss(av)}}$ is the average value of the motor losses and R_{th} is the published value of the thermal resistance R_{th} in °C/W. A large ripple can obviously lead to overheating if the average loss is large enough by itself to raise the average winding temperature by 110°C. At an ambient temperature of 40°, the maximum winding temperature of 150° would be exceeded by $\Theta_{pk} - \Theta_{av}$.

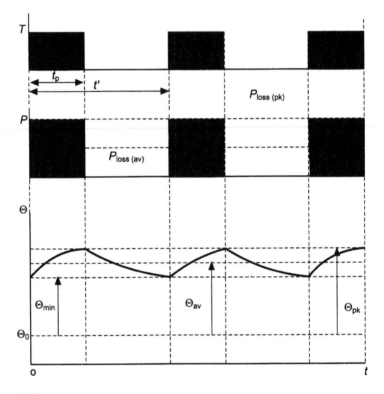

Figure 5.3
Effect of intermittent torque on the motor winding temperature

The rate of rise of 'the motor temperature' from ambient to steady state depends on the thermal time constant τ_{th}. For a typical motor, τ_{th} is of the order of 35 minutes. The published value is normally the overall time constant of the main mass, which for the brushless motor is taken to be the stator winding, stator iron and motor case. The temperatures of these three motor components do not vary at the same rate. The $i^2 R$ heating energy passes from the winding and through the stator core before reaching the motor case of the motor, and so the winding must heat up before the stator core. The winding heats quickly, particularly at the start of the torque pulse, and also cools quickly at the start of the pulse gap. This means that the average time constants of the rising   and falling waveforms of winding temperature in Figure 5.3 are lower than the value of τ_{th} quoted for the motor as a whole [9].

The ripple on the winding temperature becomes more pronounced as t' is increased. The ripple also increases as t_p is reduced, assuming the average losses are kept the same by increasing the torque. The extreme case would be if the motor produced four times its continuous rated torque over a pulse width of $t_p = t'/16$. As a rule of thumb, the ripple in the steady-state winding temperature will normally be confined to a band of $\pm 10°C$ when

$$\tau_{th} > 100t'$$

where $\tau_{th} > 25$ minutes and the maximum average temperature rise is $100°C$. When $\tau_{th} < 100t'$, the ripple may still fall within the $\pm 10°C$ band at the higher end of the tp/t' range. Such cases should be considered individually.

Calculations are relatively easy when the period of the intermittent operation is less than 1% of the thermal time constant of the motor. The maximum ripple is $\pm 10°C$. The designed maximum rise in the winding temperature is $110°C$, and so the maximum average losses are those which lead to an average winding temperature of $110 - 10$, or $100°C$ above ambient. The losses are found by adding the i^2R loss to the speed-sensitive loss as estimated from the Soac diagram.

Example 5.2

The sinusoidal motor of Example 5.1 is to provide the intermittent torque shown in Figure 5.4 at a speed of 2500 rpm. What is the maximum height T_{max} of the torque pulse, *assuming that the peak temperature rise of the winding is limited to 110°C? The motor constants are as follows:*

$$R_{LL} = 1.29 \ \Omega$$
$$K_T = 0.51 \ Nm/A$$
$$\tau_{th} = 35 \ minutes$$
$$R_{th} = 0.75° C/W$$

The current at T_{max} is $I = \dfrac{T_{max}}{K_T} = \dfrac{T_{max}}{0.51}$ A.

The effective stator resistance for the purpose of calculation of the i^2R loss in the sinusoidal motor is $1.5 \ R_{LL}$. The effective stator resistance at the average, steady-state temperature is

$$R' = 1.5 \ R_{LL}[1 + 0.00385(\Theta_{ss} - 25)] \ \Omega$$

The ratio of the thermal time constant of the motor to the period of the intermittent operation is $35 \times 60/18$, or 117. We may assume that the maximum ripple will be within $\pm 10°C$, and that the average temperature of the winding can be allowed to rise by $110 - 10$, or $100°C$. In the extreme case, the average winding temperature would then be $140°C$ at an ambient of $40°C$. The maximum, average, effective stator resistance is therefore

$$R' = 1.5 \times 1.29[1 + 0.00385(140 - 25)] = 2.79 \ \Omega$$

The i^2R loss in watts in the stator winding is

$$I^2R' = (T_{max}/0.51)^2 \times 2.79$$

or

$$I^2R' = 10.73 \ T^2_{max}$$

Following the approach used in Example 5.1 gives the speed-sensitive loss at 2500 rpm as

$$P_{sp} \approx 20 \ W$$

The total peak and average losses are

$$P_{loss(pk)} = (10.73 T^2_{max} + 20)$$

and

$$P_{\text{loss(av)}} = \frac{t_p}{t'} P_{\text{loss(pk)}} = (1.19 T_{\text{max}}^2 + 2)$$

where $t' = 9t_p$. The average temperature rise of the winding is

$$\Theta_{\text{av}} = R_{\text{th}} P_{\text{loss(av)}}$$

or

$$100 = 0.75(1.19 T_{\text{max}}^2 + 2)^\circ\text{C}$$

giving

$$T_{\text{max}} = 10.5\,\text{Nm}$$

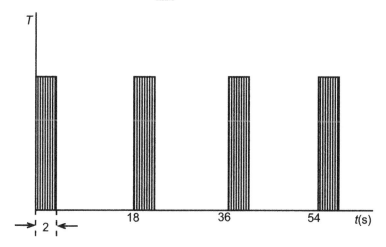

Figure 5.4
Motor torque for Example 5.2

Tolerances

The above figures are based on the nominal values of torque constant and winding resistance. In an extreme case, where K_T is below and R above the nominal value by 10%, the maximum torque would be

$$T_{\text{max}} = 9.0\,\text{Nm}$$

Motor tests

The motor in the above example was tested on the intermittent duty cycle shown in Figure 5.5 under the following conditions:

$$T = 10.5\,\text{Nm}$$
$$t' = 9t_p$$
$$\tau_{th} = 78t'$$
$$\Theta_0 = 29°C$$

The ratio t'/t_p was equal to 9 for the both the test and Example 5.2. The test was, however, made pessimistic by reducing the ratio τ_{th}/t' from the figure of 117 used for the example to 78, well below the recommended minimum of 100. When the temperature of the test motor had reached steady-state conditions, the variation in the winding temperature was recorded. The results are shown in Table 5.1, together with the nominal figures predicted in Example 5.2. At the ambient temperature of 29°C, the average rise of the test motor was 97°C, with a ripple of $\pm 4°C$. Correcting the figures to allow for the maximum ambient temperature of 40°C gives a rise for the motor of $100 \pm 5°C$. The test motor had nominal values of resistance and torque constant, and so its average rise in temperature may be expected to be in close agreement with the value used in the example.

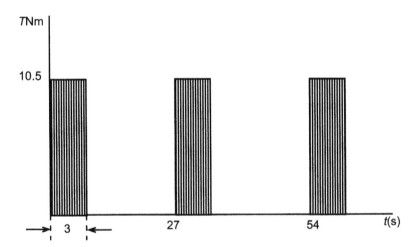

Figure 5.5
Torque cycle for motor test

Table 5.1 Steady-state temperature ripples

Temperature rise ^oC	Example 5.2 ($\theta_o = 40^oC$)	Test ($\theta_o = 29^oC$)
Θ_{pk}	110	101
Θ_{av}	100	97
Θ_{min}	90	93

Thermal models

In Example 5.2 and in the motor test, the demand was known in terms of an intermittent torque. In practice there may be other intermittent effects which are more difficult to analyse. An extreme example would be where pulses of the maximum allowable motor current occur when the motor is at standstill. Thermal models are useful in such cases, where, for example, the thermal resistance of the motor is split into the winding-to-case and case-to-ambient values, and the thermal capacity into the winding and case capacities [9].

5.5 Incremental motion

The aim of this section is to assist in the selection of a motor for applications where the load is to be driven incrementally over a specified load velocity profile, and where the load and motor inertias should if possible be matched. Calculations are based on the i^2R loss, plus an allowance for the speed-sensitive loss.

It was shown in Section 4.5 that the trapezoidal load velocity profile is most efficient when the periods of acceleration, deceleration and constant speed are all equal. Such a profile may not be realizable in practice, and the three periods may differ from each other. In this section we take an approximate account of the profile shown in Figure 4.16 and repeated here in Figure 5.6, where the load accelerates and decelerates over the proportions p_1 and p_2 of the motion time t_p. The total period of the increment is t'.

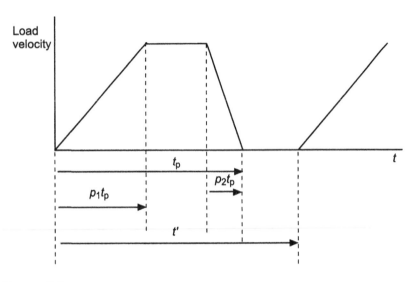

Figure 5.6
Trapezoidal profile of load velocity ($p_1 + p_2 < 1$)

Motor selection for the rotating load

The principles of optimization were introduced in Chapter 4 on the basis of the motor-load dynamics. The thermal ratings of the motors used in the examples were assumed to be high enough to cope with demands of the given loads. This section describes a method of motor selection based on the thermal rating needed for a particular load.

Stator heating energy

In Section 4.5, the energy produced in the form of heat by the stator $i^2 R$ loss was found for the case of a load consisting of a rotating mass of inertia J_L, and an opposing torque T_L. When the load rotates through θ_p in time t_p, the energy is

$$\epsilon = \frac{Rt_p}{K_T^2} \left[c_p \left(\frac{JG\theta_p}{t_p^2} \right)^2 + \left(\frac{T_L}{G} \right)^2 \right]$$

where $J = (J_m + J_L/G^2)$ and G is equal to unity when the load is driven directly from the motor. For the purpose of optimization, we took account of any trapezoidal profile by using a general expression for the profile constant c_p. The constant takes its minimum value of 13.5 when the trapezoidal profile of load velocity is equally distributed. In this section, however, we are interested in an approximate method of motor selection and so the exact value of the profile constant is not important. To allow for some asymmetry we will use a value of 15 in the initial selection of the motor. When $T_L = 0$, the stator heating energy is therefore

$$\epsilon \approx \frac{15 R J^2 G^2 \theta_p^2}{K_T^2 t_p^3}$$

Selection criterion

For matched motor and load inertias, and zero load torque, the square of the optimum reducer ratio has been shown to be

$$G_0^2 = \frac{J_L}{J_m}$$

and so

$$J^2 G_0^2 = \left(J_m + \frac{J_L}{G_0^2} \right)^2 G_0^2 = 4 J_m J_L$$

For the matched inertia case, we may therefore substitute $4 J_m J_L$ for $J^2 G^2$ in the last expression above for ϵ. At the maximum allowable motor temperature of 150°C, the steady-state $i^2 R$ power loss for the matched inertia case becomes

$$P_{\text{loss(max)}} = \epsilon/t' \cong \frac{60 R' J_m J_L \theta_p^2}{K_T^2 t_p^3 t'}$$

where $R' = 1.48 R$. For most incremental motion applications, $\tau_{\text{th}} \gg t'$ and any heating variations will be filtered out. The maximum allowable rise in motor temperature is 110°C, and so

$$P_{\text{loss(max)}} < \frac{110}{R_{\text{th}}}$$

The mechanical time constant of the motor is

$$\tau_m = \frac{RJ_m}{K_T K_E}$$

For the trapezoidal motor, $K_E = K_T$. Combining the last three equations above gives the limiting condition as

$$R_{th}\tau_m < 1.2\frac{t_p^3 t'}{J_L \theta_p^2}$$

Sinusoidal and trapezoidal motors have much the same mechanical time constant, thermal resistance and rating. The last expression should therefore apply to both forms. So far we have dealt with only the $i^2 R$ loss for the inertial part of the load. When the load includes an opposing torque T_L, it can be shown that the $i^2 R$ loss increases by the factor

$$\delta_1 = 0.5[1 + \sqrt{(1+\Delta_1)}] \quad \text{where} \quad \Delta_1 = \frac{1}{15}\left(\frac{T_L t_p^2}{\theta_p J_L}\right)^2$$

Approximately 50% more may be added to cover the speed-sensitive loss, although this figure is very pessimistic at the lower end of the speed range. When we take account of all the $i^2 R$ and speed-sensitive losses, the motor selection criterion becomes

$$R_{th}\tau_m < \frac{0.8 t_p^3 t'}{J_L \theta_p^2 \delta_1}$$

Rating coefficient

Table 5.2 lists the values of motor constants given in typical specification sheets and also the values of the rating coefficient $R_{th}\tau_m$ °C ms/W, for a range of sinusoidal motors which increase in size down the table. K_T is given as the total torque constant for the three phases. Four different voltage specifications are available for each motor shown on the table, but the choice of voltage does not significantly affect the rating. The speed limits are set by dividing the no-load voltage per 1000 rpm gradient by the peak supply voltage for the particular motor chosen for each line of the table.

Table 5.2 Constants for various sinusoidal motors

Motor	K_T	$J_m \times 10^3$	R_{LL}	$\tau_m \times 10^3$	R_{th}	$R_{th}\tau_m \times 10^3$
01	0.75	0.05	37	8.8	1.7	15.0
02	0.75	0.085	10.5	3.5	1.36	4.8
03	0.75	0.15	3.4	1.8	0.96	1.7
04	1.02	0.18	5.0	1.6	0.80	1.3
05	1.53	0.22	7.6	1.4	0.77	1.08
06	1.53	0.24	11.4	2.1	0.75	1.6
07	1.53	0.35	6.4	1.6	0.71	1.14
08	1.53	0.46	3.7	1.3	0.68	0.88
09	1.53	0.68	2.1	1.0	0.60	0.60
10	1.53	1.2	1.7	1.4	0.55	0.77
11	2.10	1.7	1.9	1.1	0.52	0.57
12	3.03	5.6	2.5	2.3	0.34	0.78
13	3.03	8.2	1.22	1.8	0.30	0.54
14	4.20	11.0	1.5	1.5	0.27	0.41
15	4.20	14.0	1.13	1.4	0.23	0.32

Speed limits: M01–04 = 6000 rpm, M05–10 = 5400 rpm, M11 = 3900 rpm, M12–13 = 2700 rpm, M14–15 = 1900 rpm.

The values of $R_{th}\tau_m$ are directly affected by the value of τ_m and do not fall smoothly as the size of the motor increases. Selection may be based on the combination of the rating coefficient with the higher or lower value of time constant as required. In the following examples it is assumed that the motor is to be chosen solely on the basis of its rating coefficient. Following an initial selection, the rms torque required from the prospective motor is calculated for both the nominal and extreme values of K_T and R. The point of selection may then need to be adjusted to the next larger or the next smaller motor.

Example 5.4

A load is to be rotated incrementally over the velocity profile shown in Figure 5.7. The loading constants are:

$$\theta_p = 2\pi\,rad$$
$$t' = 0.20\,s$$

$$t_p = 0.12\,s$$
$$J_L = 0.0082\,kgm^2$$
$$T_L = 20\,N$$

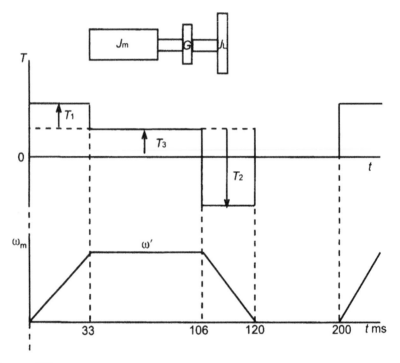

Figure 5.7
Motor torque and velocity for Example 5.4

Motor selection

The load torque factor in the denominator of the rating criterion above is

$$\delta_1 = 0.5[1 + \sqrt{(1 + \Delta_1)}] \quad \text{where} \quad \Delta_1 = \frac{1}{15}\left(\frac{T_L t_p^2}{\theta_p J_L}\right)^2$$

Inserting the numerical values gives $\delta_1 = 1.4$. An initial choice is made from Table 5.2 at the maximum value of $R_{th}T_m$ which satisfies

$$R_{th}T_m < \frac{0.8 \times 0.12^3 \times 0.2}{0.0082 \times (2\pi)^2 \times 1.4}$$

or

$$R_{th}\tau_m < 0.61 \times 10^{-3}$$

The smallest motor in Table 5.2 with a value of $R_{th}\tau_m$ below 0.61 is M09.

Reducer ratio

Having identified a motor, the next step is to calculate the optimum ratio of a reducer. Figure 5.7 shows the motor speed profile as the motor accelerates, runs at constant speed and then decelerates. Using the method described in Chapter 4, the optimum ratio of a reducer between M09 and the load is found to be $G_\Delta = 4.4$.

Motor speed

The constant speed of the load is

$$\omega_c = \frac{\theta_p}{t_p[1 - 0.5(p_1 + p_2)]} \quad \text{(Section 4.5)}$$

$$= \frac{2\pi}{0.12\left[1 - 0.5\left(\frac{33}{120} + \frac{14}{120}\right)\right]} = 65.1\,\text{rad/s}$$

The constant motor speed in Figure 5.7 is therefore

$$\omega' = G_\Delta\omega_c = 286\,\text{rad/s} = 2735\,\text{rpm}$$

Speed-sensitive loss

A pessimistic estimate of the speed-sensitive loss is obtained by taking ω' to be the motor speed throughout the incremental time t_p. Figure 5.2 shows the Soac curve for motor M09. Using the method followed in Example 5.1 gives the loss at 2735 rpm as

$$P_{sp} \approx 70\,\text{W}$$

The rms torque

The torque profile is shown in Figure 5.7. The torque required to accelerate and decelerate the motor and load masses is

$$T = \left(J_m + \frac{J_L}{G_\Delta^2} \right) \frac{\mathrm{d}}{\mathrm{d}t} \omega_m$$

giving

$$T_1 = \left(0.00068 + \frac{0.0082}{4.4^2} \right) \times \frac{286}{0.033} = 9.6 \, \text{Nm}$$

The same calculation for the period of deceleration gives

$$T_2 = -22.5 \, \text{Nm}$$

During the full period of motion

$$T_3 = \frac{T_L}{G_\Delta} + \frac{P_s}{\omega'} = \frac{20}{4.4} + \frac{70}{286} = 4.8 \, \text{Nm}$$

The total torques required during acceleration and deceleration are

$$T_3 + T_1 = 14.4 \, \text{Nm}$$

$$T_3 + T_2 = -17.7 \, \text{Nm}$$

The rms torque required over one duty cycle is

$$T_{\text{rms}} = \sqrt{\frac{14.4^2 \times 0.033 + 4.8^2 \times 0.073 + (-17.7)^2 \times 0.014}{0.200}}$$

$$= 8.0 \, \text{Nm}$$

The Soac curve for M09 gives the maximum continuous torque available at 2735 rpm as 8.2 Nm. This means that the motor temperature should be below its limit of 150°C at the rms torque of 8.0 Nm required in this case.

Motor temperature

The final temperature rise is

$$(\Theta_{ss} - \Theta_0) = R_{th} P_{\text{loss(av)}}$$

or

$$(\Theta_{ss} - \Theta_0) = R_{th}[I_{rms}^2 R(1 + 0.00385(\Theta_{ss} - 25) + P_{sp}\frac{t_p}{t'}]$$

Rearranging the last equation above gives the final temperature as

$$\Theta_{ss} = \frac{\Theta_0 + R_{th}\left(0.904I_{rms}^2 R + P_{sp}\frac{t_p}{t'}\right)}{1 - 0.00385R_{th}I_{rms}^2 R}$$

where $I_{rms} = T_{rms}/K_T$. For M09, $K_T = 1.53$ and so $I_{rms} = 8.0/1.53 = 5.23$ A. The final temperature of M09 at an ambient temperature of 40°C is therefore

$$\Theta_{ss} \approx \frac{40 + 0.60[0.904 \times 5.23^2 \times (2.1 \times 1.5) + 70 \times 0.12/0.20]}{1 - 0.00385 \times 0.60 \times 5.23^2 \times (2.1 \times 1.5)}$$

or

$$\Theta_{ss} \approx 140°C$$

Torque ratings

The above calculations are used when the final temperature of the motor is required. However, the suitability of the motor selected from Table 5.2 may be checked without reference to temperature by using the torque ratings given by the Soac curve. The maximum i^2R loss which can be allowed at a given motor speed is

$$I^2R = \left(\frac{T_{soac}}{K_T}\right)^2 R$$

where K_T and R are nominal values and T_{soac} is the rated torque at the given speed, as shown on the Soac diagram. If K_T and R were to take values at the extreme tolerance of 10%, the i^2R loss at the required torque T_{rms} would be

$$\left(\frac{T_{rms}}{0.9K_T}\right)^2 \times 1.1R = 1.36\left(\frac{T_{rms}}{K_T}\right)^2 R$$

The loss in the extreme case must still be below the maximum loss allowable. A safety margin is therefore defined when

$$\frac{T_{soac}}{T_{rms}} > 1.17$$

Table 5.3 shows the ratio of the nominal rated torque to the torque demand for motors M10, 11, 12 and 13, at the motor speeds required when driving the load specified in the present example. The rms torque required from each motor has been found by the method followed above for M09, using speed-sensitive losses calculated from the Soac curves (not shown) of 45 W, 50 W, 55 W and 50 W respectively. The nominal, final temperature of M09 has already been predicted to be 140°C, which correlates with the figure of 1.03 for the T_{soac}/T_{rms} ratio. M11 has a ratio of 1.14, only slightly below the figure of 1.17 required for the full safety margin. The rating of M13 has a good safety margin for the load specified in the present example.

Table 5.3   Motor torques and ratings for Example 5.4

	M09	*M10*	*M11*	*M12*	*M13*
Nominal T_{soac}	8.2	10.0	14.4	22.0	33.5
Required T_{rms}	8.0	12.0	12.6	23.0	27.6
T_{soac}/T_{rms}	1.03	0.83	1.14	0.96	1.21

Motor selection for the translating load

Following the approach used for the rotating load, the motor selection criterion for a load consisting of mass m which moves distance x in time t_p against force F is found to be

$$R_{th}T_m < \frac{0.8t_p^3 t'}{mx^2\delta_2}$$

where $\delta_2 = 0.5[1 + \sqrt{(1 + \Delta_2)}]$ and $\Delta_2 = \frac{1}{15}\left(\frac{Ft_p^2}{xm}\right)^2$.

The belt and pulley drive

Example 5.5

Select a motor from Table 5.2 for the following belt driven load:

$$x = 20\,cm$$
$$t' = 0.20\,s$$
$$t_p = 0.05\,s$$
$$c_p = 13.5$$
$$m = 0.4\,kg$$
$$F = 10\,N$$

Inserting the numerical values gives

$$\Delta_2 = \frac{1}{15}\left(\frac{10 \times 0.05^2}{0.20^2 \times 0.4}\right) = 0.0065$$

The value of Δ_2 is too low to affect the motor rating coefficient. An initial selection is made where

$$R_{th}\tau_m < \frac{0.8 \times 0.05^3 \times 0.20}{0.4 \times 0.20^2} = 1.25 \times 10^{-3}$$

From Table 5.2, we see that M05 should be suitable. The radius of the drive pulley can now be optimized (Section 4.5). Allowing $J' = 0.0001$ kg m² for the approximate inertia of a small pulley, the optimum radius is found to be $r_0 = 2.83$ cm.

The motor velocity and torque profiles are shown in Figure 5.8. The load travels 20 cm as the motor shaft turns through $\theta_m = 20/2.83 = 7.07$ rad. The velocity profile is distributed symmetrically, and so the motor speed over the flat-topped section is

$$\omega' = 3\theta_m/2t_p = 3 \times 7.07/(2 \times 0.05) = 212\,rad/s = 2025\,rpm$$

The inertia of the motor and drive pulley is matched with the inertia of the belt and load. The torque required for acceleration and deceleration is

$$T_1 = -T_2 = 2(J_m + J')\frac{d}{dt}\omega_m = 2 \times 0.00032 \times \frac{212}{0.017} = 8.0\,Nm$$

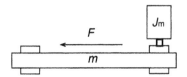

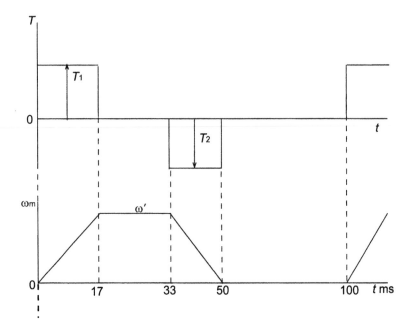

Figure 5.8
Motor torque and velocity for Example 5.5

The speed-sensitive loss for M05 at 2025 rpm is estimated to be 30 W (Soac curve not shown). The motor torque required to provide force F, and for the generation of the speed-sensitive loss, is

$$T_3 = Fr + \frac{P_{sp}}{\omega'} = 10 \times 0.028 + 30/212 = 0.4 \text{ Nm}$$

The rms torque over the duty cycle is therefore

$$T_{rms} = \sqrt{\frac{8.4^2 \times 0.017 + 7.6^2 \times 0.017}{0.20}} = 3.3 \text{ Nm}$$

Note that the torque of 0.4 Nm during the constant speed

period may be neglected in the calculation of the rms torque. The Soac curve for M05 shows the nominal torque continuously available at 2025 rpm to be 4.1 Nm, and so the maximum torque available at the safety margin is $4.1/1.17 = 3.5$ Nm. This figure is above the torque required and so motor M05 would not overheat, whatever the actual values of torque constant and winding resistance within the $\pm 10\%$ band.

Table 5.4 shows the results when the above calculations are made for the next two smaller motors in Table 5.2, using speed-sensitive losses of 30 and 25 W respectively. The table shows a good safety margin for the rating of the motor initially selected on the basis of its rating coefficient. The margin for M04 is too small, and M03 is unsuitable even in the nominal case.

Table 5.4 Motor torques and ratings for Example 5.5

	M05	M04	M03
Nominal T_{soac}	4.1	3.4	2.7
Required T_{rms}	3.3	3.1	2.9
T_{soac}/T_{rms}	1.24	1.10	0.93

The ball screw drive

In Chapter 4 the pitch of the ball screw was optimized for any drive motor, and the rating of the motor used in Example 4.3 was not considered. We can now check if the motor used in the example was the most suitable for the given load, on the basis of the motor rating.

Example 5.6

Select a correctly rated motor for the load of Example 4.3. The load constants are

$$x = 0.025\,m$$
$$t_p = 0.120\,s$$

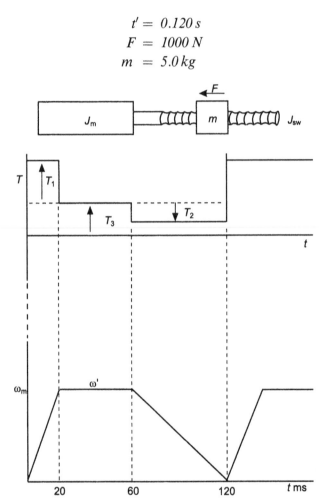

Figure 5.9
Motor torque and velocity for Example 5.6

Figure 5.9 shows the motor torque and velocity profiles. The motor should satisfy

$$R_{th} T_m < \frac{0.8 t_p^3 t'}{m x^2 \delta_2}$$

where

$$\delta_2 = 0.5[1 + \sqrt{(1 + \Delta_2)}] \quad \text{and} \quad \Delta_2 = \frac{1}{15} \left(\frac{F t_p^2}{x m} \right)^2$$

Inserting the numerical values gives $\Delta_2 = 885$, and $\delta_2 = 15.4$. The motor rating coefficient is found to be

$$R_{\text{th}} \tau_{\text{m}} < 3.4 \times 10^{-3}$$

Table 5.2 gives M03 as the initial choice. Following the method used in Example 4.3, using $c_p = 18$ and assuming a screw inertia of $J_{\text{sw}} = 0.00003$ kgm^2, gives the optimum pitch of a ball screw driven by M03 as $d_\Delta = 7.2$ mm.

The load moves 25 mm as the screw and motor shaft turn through $25/7.2 = 3.47$ revolutions in 0.12 s. The motor shaft rotates through $\theta_{\text{m}} = 3.47 \times 2\pi = 21.8$ rad, and the motor speed during the flat-topped section of the velocity profile of Figure 5.9 is

$$\omega' = \frac{\theta_{\text{m}}}{t_{\text{p}}[1 - 0.5(p_1 + p_2)]}$$

giving

$$\omega' = 273 \,\text{rad/s} = 2607 \,\text{rpm}$$

The inertia of mass m reflected to the motor side of the screw is $md^2/4\pi^2$ (Section 4.5). The motor torque required for acceleration or deceleration of the motor, screw and load masses in Figure 5.9 is

$$T = (J_{\text{m}} + J_{\text{sw}} + \frac{md^2}{4\pi^2})\frac{\text{d}}{\text{d}t}\omega_{\text{m}}$$

giving

$$T_1 = 2.55 \text{ Nm and } T_2 = -0.84 \text{ Nm}$$

The work done in supplying the load force over a distance of one screw pitch is

$$Fd = 2\pi T_0$$

where T_0 is the output torque of the motor (assuming negligible losses at the screw). The motor output torque required for the supply of the load force is therefore

$$T_0 = \frac{Fd}{2\pi}$$

The Soac diagram for M03 is shown in Figure 5.2. The speed-sensitive loss at 2600 rpm is found to be $P_{sp} \approx 30$ W. The total motor torque required for the supply of the load force is therefore

$$T_3 = \frac{Fd}{2\pi} + \frac{P_{sp}}{\omega'} = 1.26 \, \text{Nm}$$

The torques required during the three sections of the period of motion are

$$T_1 + T_3 = 3.81 \, \text{Nm}$$
$$T_3 = 1.26 \, \text{Nm}$$
$$T_2 + T_3 = 0.42 \, \text{Nm}$$

The rms value of the three torques is $T_{rms} = 1.74$ Nm.
The maximum, nominal torque is $T_{soac} = 2.65$ Nm.
The maximum safe torque in the worst case would be

$$T_{soac}/1.17 = 2.26 \, \text{Nm}.$$

Table 5.5 shows the results when the calculations are made for M02 using a speed-sensitive loss of 25 W. The motor is seen to be unsuitable, even on a nominal basis. M03 has a more than adequate safety margin, which is reflected by the amount by which its rating coefficient falls within the limit set by the loading conditions of Example 4.3. M05, the original motor chosen at random (for example 4.3) would clearly be much too large.

Table 5.5 Motor torques and ratings for Example 5.4

	M03	M02
Nominal T_{soac}	2.65	1.28
Required T_{rms}	1.74	1.38
T_{soac}/T_{rms}	1.52	0.93

5.5 Precautions

The temperature of 150°C is the absolute maximum for any part of a brushless servomotor. Overheating of the motor or its environment is an obvious risk if the safety margin used in selecting the motor is too small. On the other hand, an overcautious approach can result in an unnecessarily large and expensive motor. The final choice is often influenced by several factors, and no figure can be given for a safety margin which suits all circumstances.

Throughout this chapter we have allowed for 10% variation of the stator resistance and torque constant, between individual motors of the same type. We have catered for this variation by allowing a margin of 17% between the required rms torque and the rated torque of the motor. It is in fact possible for the torque constant to fall by 10% as the temperature rises, mainly due to the fall in the magnetic intensity of the permanent magnets. The same allowance for the stator resistance is, however, very generous. The normal variation is much smaller, being dependent, for example, on very small differences between the nominal and actual diameter of the wire used in the manufacture of the winding. The conclusion is that the margin of 17% between the required and rated torque should be adequate, assuming that there is no other reason why the motor should be derated.

In practice there may be several thermal effects to be taken into account, apart from the effect of the tolerance allowed in the values of the motor constants. For example, an excessive amount of heat transmitted directly from a hot motor may cause distortion of the frame of a precision mechanism such as a ball screw. It is also important that the overall design of any system should allow for sufficient space between the hot surface of the motor and any heat-sensitive equipment, or allow for the accommodation of a larger motor which can run at a lower temperature. Care is also required in the way the motor is fitted to the frame of the installation. As already mentioned in Section 5.2, the thermal resistance of a motor is likely to be higher than the published value when it is thermally isolated from the frame. Thermal isolation can be deliberate, or it can be the accidental result of the insertion of electrical insulation material at all points of metal to metal contact with the frame. The maximum rms torque available in either case will be lower than the figure given by the Soac curve, which is plotted experimentally when the motor is bolted directly to a typical frame. If possible, a heatsink should be fitted whenever the frame has to be electrically or thermally isolated from the motor.

Assuming that the motor has been selected correctly on the basis of its ability to supply the required torque, and that all other factors have been taken into account, any overheating which does occur is normally the result of accidental misuse. Computer controlled duty cycles are very easy to change, and a motor which has been correctly rated for a particular torque profile is likely to overheat if the profile is made more demanding. It is, of course, sometimes difficult to judge the range of the future demands of an application. Where space permits, the fitting of a blower will help an existing motor to cope with a moderate increase in demand.

References

1. Miller, T. J. E. (1989). *Brushless Permanent-Magnet and Reluctance Motor Drives*. Clarendon Press.
2. Hendershot, J. R. Jr. and Miller, T. J. E. (1994). *Design of Permanent-Magnet Motors*. Magna Physics Publishing and Clarendon Press.
3. Dote, Y. (1990). *Brushless Servomotors: Fundamentals and Applications*. Clarendon Press.
4. Electro-craft Corp. (1980). DC Motors, speed controls, servo systems.
5. Armstrong, R. W. Jr. (1998). *Load to Motor Inertia Mismatch: Unveiling the Truth*. Drives and Controls Conference, Session 6, 17–22.
6. Newall, P. (1998a). *Models of the Motor and Load: Analytical Solutions*. Paper DD0369, SEM Ltd.
7. Newall, P. (1998b). *Motor and Load Mechanical Resonance*. Paper DD326, SEM Ltd.
8. Newall, P. (1998c). *Models of the Motor and Load: Numerical Solutions*. Paper DD0370, SEM Ltd.
9. Newall, P. (1996). *Brushless Motor Thermal Models*. Paper DDO230, SEM Ltd.

Index

Printed and bound by CPI Group (UK) Ltd, Croydon, CR0 4YY

03/10/2024

01040433-0020